CRYSTAL ENGINEERING
A TEXTBOOK

CRYSTAL ENGINEERING
A TEXTBOOK

Gautam R Desiraju

Indian Institute of Science

Jagadese J Vittal

National University of Singapore

Arunachalam Ramanan

Indian Institute of Technology Delhi

IISc
Press

World Scientific

NEW JERSEY · LONDON · SINGAPORE · BEIJING · SHANGHAI · HONG KONG · TAIPEI · CHENNAI

Published by

World Scientific Publishing Co. Pte. Ltd.

5 Toh Tuck Link, Singapore 596224

USA office: 27 Warren Street, Suite 401-402, Hackensack, NJ 07601

UK office: 57 Shelton Street, Covent Garden, London WC2H 9HE

British Library Cataloguing-in-Publication Data
A catalogue record for this book is available from the British Library.

CRYSTAL ENGINEERING
A Textbook

ISBN-13 978-981-4338-75-2
ISBN-10 981-4338-75-3
ISBN-13 978-981-4366-86-1 (pbk)
ISBN-10 981-4366-86-2 (pbk)

Typeset by Stallion Press
Email: enquiries@stallionpress.com

Printed in Singapore.

Contents

Preface

The fascinating subject of crystal engineering stands at the crossroads of chemistry and crystallography. The term *crystal engineering* was first suggested in 1955. Active research in this area has been ongoing for the past 25 years. This textbook is the first that is devoted exclusively to this interdisciplinary subject.

The division of chemistry into organic, inorganic and physical branches is traditional in teaching programs. For the undergraduate student who wishes to embark on a research career in these core areas, there is a natural continuity between what is learnt in the classroom, and what is done in the research laboratory. However, for those who enter research in areas that do not fall neatly into one of the above streams there is often little connection between one's experience as an undergraduate student and then subsequently as a research scholar. It is believed that the student will somehow acquire the basic skills needed to do research in a subject like crystal engineering from what he or she has learned in courses in organic, inorganic and physical chemistry. This is not always possible and certainly not easy.

Crystallography as a subject is insufficiently handled in course curricula, considering its pervasive influence in chemistry. With the proliferation of automated table top diffractometers and black box software, the determination of crystal structures of organic and inorganic substances seems to have become routine and automated. Challenges in small molecule crystallography today have more to do with understanding the reasons why a compound adopts a particular crystal structure and not another. Crystal engineering is part of such an endeavour because it attempts to establish packing trends in whole families of compounds. The use and relevance of crystallography in crystal engineering is much more extended than its application in traditional organic and inorganic chemistry where it is used generally only to confirm and establish chemical structure. These wider implications of crystallography as something that goes beyond crystal structure determination do not generally find a place in undergraduate teaching programs.

This textbook is intended to fill these gaps in chemistry and crystallography teaching. It is meant for senior undergraduates and entry level PhD students who are interested in structural chemistry and crystal engineering. It can easily be used as a text for an elective course of say 30 to 40 hours in the final stages prior to a PhD program. An understanding of crystal engineering requires some appreciation of crystal structures. It is helpful but not mandatory to have some familiarity with X-ray crystallography, crystal structure analysis and data retrieval methods from crystallographic databases. We have not attempted to get into a detailed treatment of any of these topics because there are already several such books available. It is also easy to become familiar with these matters from information from the internet.

The use of this book will be greatly enhanced if the student has access to the Cambridge Structural Database (CSD). This is the primary tool for mining and retrieval of crystallographic information pertaining to organic, organometallic and metal-organic structures. No meaningful study of crystal engineering is possible without the CSD and it is suggested that all serious students of the subject attempt to gain access to this software as quickly as possible.

The structuring of this book roughly corresponds to the chronology of research developments, except for the fact that crystal engineering of organic and metal-organic compounds have proceeded simultaneously and independently during the last two decades. More recently, many have appreciated that both these major areas have much to share with each other: it is easy to club organic crystal engineering and the study of coordination polymers together into the single subject of *crystal engineering*.

The teacher should use this text as a starting point, and not hesitate to consult the literature mentioned at the end of each chapter to provide students a more advanced treatment of selected topics. This reading list does not claim to be exhaustive. There is a preference for earlier papers that introduce new concepts and ideas. The current literature in the subject is exploding and is well covered in two scientific journals, *Crystal Growth and Design* from the American Chemical Society and *CrystEngComm* from the Royal Society of Chemistry. It is hoped that in this process of further reading, the teacher who is not engaged in research in crystal engineering will also be tempted to start research in the area. The philosophical differences between a physical organic chemist or a synthetic chemist and the organic crystal engineer are actually very slight. Similarly, inorganic chemists should be able to easily recast themselves as crystal engineers who work in the field of coordination polymers. Computational chemists are now seeking new pastures in problems pertaining to supramolecular structure; to do computational crystal engineering is a natural progression in such an attempt.

Crystal engineering is a subject that seeks to establish connections between structure and function. It is therefore of very wide scope and we have observed that many PhD students in this area have been able to establish themselves meaningfully in their professional lives in areas that seem to be somewhat removed from crystal engineering itself. These areas include materials science and engineering, pharmaceutical science and a wide variety of technological enterprises. This adaptability is inherent in any subject that lies at the intersection of many thought streams. Students come to this subject with different viewpoints, and take away from it different lessons that they have learned. It is hoped therefore that this textbook will be read with the open spirit in which it was written.

G. R. Desiraju, J. J. Vittal, A. Ramanan

Bangalore, Singapore, New Delhi
June, 2011

Acknowledgements

The authors would like to thank the following for permission to reproduce photographs and illustrations: Michael J. Zaworotko (University of South Florida), T. N. Guru Row (Indian Institute of Science), Istvan Hargittai (Hungarian Academy of Sciences), Ulrich Griesser (Universität Innsbruck), William Jones (University of Cambridge), Paola Spadon (Università degli Studi di Padova), Lian Yu (University of Wisconsin), Kumar Biradha (IIT-Kharagpur), Bart Kahr (New York University), J. Stephen Clark (University of Glasgow), Stuart Batten (Monash University), Wenbin Lin (The University of North Carolina), Joachim Ulrich (Universität Halle) and Matthias Forster (Stiftung Preußische Schlösser und Gärten Berlin-Brandenburg).

Figures and tables taken from journals of the following societies have been reproduced with permission: American Chemical Society, Royal Society of Chemistry, Elsevier, International Union of Crystallography, Oxford University Press, International Union of Pure and Applied Chemistry. Illustrations have been reproduced with permission from the National Portrait Gallery, London and from the Oregon State University Library (Ava Helen and Linus Pauling Papers).

We are grateful to our colleagues and friends in our respective organizations, Indian Institute of Science, National University of Singapore and Indian Institute of Technology Delhi for sharing their teaching experiences, reading portions of the manuscript and their general comments and suggestions. A number of students and post-doctoral associates contributed to the text as it took shape. Several of them also assisted in the preparation of illustrations and problems. These students and post-doctorals include, but are not restricted to the following individuals: Tejender Singh Thakur, Arijit Mukherjee, Srinu Tothadi (Bangalore); Goutam Kumar Kole, Mohammad Hedayetullah Mir, Mangayakarasi Nagarathinam (Singapore); Monika Singh, Dinesh Kumar (New Delhi).

Finally, we would like to record the support and encouragement received from our respective families during this project.

Copyright Permissions

The following images have been reprinted with permission from the various sources.

Chapter 1

Page 4, School of Chemistry, University of Glasgow.
Page 12, National Portrait Gallery, London

Chapter 2

Page 33, Oregon State University Library (Ava Helen and Linus Pauling Papers).

Chapter 4

Page 78, Figs. 22(a) and (b), *Chemical Reviews 101*, **893**, 2001. Copyright (2001) American Chemical Society.
Page 79, Fig. 1, *Crystal Growth & Design 10*, **1866**, 2010. Copyright (2010) American Chemical Society.
Page 80, Fig. 3.1, PhD thesis of Dr. Heinrich entitled "Determination of crystallization kinetics using *in situ* measurement techniques and model-based experimental design & analysis", Zentrum für Ingenieurwissenschaften, Verfahrenstechnik/TVT, D-06099 Halle, Germany.
Page 81, Fig. 2, *Acta Crystallographica B 62*, **341**, 2006. Copyright (2006) International Union of Crystallography and Oxford University Press.
Page 82, Fig. 1, *Journal of Crystal Growth 314*, **163**, 2011. Copyright (2011) Elsevier (Academic Press).
Page 93, *Scheme 1, Pure & Applied Chemistry 58*, **947**, 1986. Copyright (1986) International Union of Pure and Applied Chemistry.

Chapter 5

Page 112, Fig. 11, *Journal of Thermal Analysis and Calorimetry 64*, **37**, 2001. Copyright (2001) Akademiai Kiado, Budapest.

Chapter 6

Book cover of *Crystalline Molecular Complexes and Compounds Volume 1* by F. H. Herbstein. Copyright (2005) International Union of Crystallography and Oxford University Press.
Page 147, Figure 1, *International Journal of Pharmaceutics 374*, **82**, 2009. Copyright (2009) Elsevier (Academic Press).

Chapter 7

Page 156, Matthias Forster (Stiftung Preußische Schlösser und Gärten Berlin-Brandenburg).

Pages 163, 164, *Coordination Polymers, Design, Analysis and Application* by S. R. Batten, S. M. Neville & D. R. Turner, 2009, Figures 1.1(a) (page 2), 2.25(b) and (c) (page 44). Copyright (2009) Royal Society of Chemistry.

Pages 163, 164, Figs. 2(a), 2(b), 13 and 14 left *Journal of Solid State Chemistry 152*, **3**, 2000. Copyright (2005) Elsevier (Academic Press).

Page 170, Fig. 3, *Chemical Society Review 34*, **109**, 2005. Copyright (2005) Royal Society of Chemistry.

Page 182, Contents figure, *Journal of American Chemical Society 126*, **2016**, 2004. Copyright (2004) American Chemical Society.

Page 183, Fig. 1A; *Journal of American Chemical Society 130*, **11584**, 2008. Copyright (2008) American Chemical Society.

Page 172, Figs. 3(b), (f) and (h), *Nature Chemistry 2*, **838**, 2010. Copyright (2010) Nature Publishing Group.

Crystal Engineering

1

Crystal engineering is the understanding of intermolecular interactions in the context of crystal packing and the utilization of such understanding in the design of new solids with desired physical and chemical properties. It is a subject of great scope and application that has developed by a coming together of thought streams from many other subjects. During the last 30 years, it has attracted the attention and interest of a varied group of scientists, notably crystallographers and chemists. The purpose of this text book is to provide a brief, basic introduction to this fascinating and important subject that has moved from the fringes into the mainstream of chemistry. Crystal engineering is concerned primarily with molecular solids. We need to distinguish these substances from extended solids such as rocksalt, diamond and metal oxides.

The molecular concept is fundamental to chemistry. Our awareness of the *molecule* as an entity in itself originates from the time when organic chemistry became a separate subject. In 1828, Friedrich Wöhler synthesized urea from ammonium cyanate (Chapter 3) in an experiment that was counter intuitive for that time. Gradually, there arose the idea of the molecule, and during much of the 19th century this molecular paradigm became reinforced through the efforts of the legendary German chemists of that time. The molecule is a group of atoms held together with interactions that are so strong that it remains relatively stable under many variations in temperature and pressure. Molecules do not normally revert spontaneously to atoms. A molecule of, say phenol is the form of that chemical substance that exists in the gas, liquid and solid phases of the compound. The thermal energy that is needed to convert solid phenol to a liquid and eventually to a gas is much smaller than the energy that is required to break the strong interactions between the C, H and O atoms that make up the

The molecule is paramount in chemistry. But rather than talk about molecules in themselves, we discuss in this book *assemblies of molecules* and why molecules associate in specific ways. A crystal is a very precise and specific type of molecular assembly. Crystal engineering teaches us how to bring molecules together exactly as we want.

Crystals have been known to mankind since antiquity. This is a photograph of one of the oldest salt pans in the world, still in use, in Aveiro, Portugal.

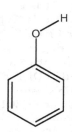

molecule of phenol. Eventually, the word *bond* was used to describe the very strong intramolecular interactions that hold atoms together in molecules and the work of Linus Pauling during the 1930s, more than anything else, gave chemists a full appreciation of the meaning of the phrase *chemical bond*.

A molecular crystal is a crystal that is made up of molecules. It is the crystalline form of any chemical substance that exists as molecules. Organic acids and bases can form cations and anions respectively. When such ions are present in a crystal, one obtains an organic salt. Such salts are also taken as molecular crystals. Not all compounds exist as molecules. Phenol, naphthalene, $SiCl_4$ and D-glucose do so but NaCl, ReO_3 and Fe_3O_4 do not. Many molecular solids are organic substances. Crystal engineering was therefore mostly synonymous with organic crystals in its initial years. During the last 20 years, however, a large number of very interesting metal–organic compounds, also called *coordination polymers*, which are highly crystalline and also of a molecular nature, have come under the scope of crystal engineering. These substances are described in Chapter 7.

A definition of the term *molecular crystal* was provided in the middle of the previous century by the great Russian physicist Alexander I. Kitaigorodskii, who said that "within a molecular crystal, it is possible to identify groups of atoms such that for every atom in a group, at least one interatomic distance within this group is significantly shorter than the smallest interatomic distance to an atom in another group." Notice that Kitaigorodskii's definition of a molecular crystal is worded in the language of geometry and not in the language of chemistry. He equates a molecule with a "group of atoms" that are defined according to distance criteria. This definition is quite useful and becomes an operational criterion of a molecular crystal. Of course, we know today that it is also possible to define a molecular crystal in chemical terms. If the energies of covalent bonds, the interactions that hold atoms together in molecules, are in the range of 75–125 kcal mol^{-1}, the energies that hold molecules together in molecular crystals are much less, by more than an order of magnitude. In Chapter 2, we will learn about the *intermolecular interactions* that hold molecules together in crystals. The energies of these interactions range between 1 and say, 20 kcal mol^{-1}. At the lower end, these energies are a little greater than the thermal energy, kT, at ambient conditions. So we can take home the idea of the molecular crystal as a collection of molecules, entities that are held internally by rather strong interactions but associated with each other by somewhat weaker interactions.

Why do we study crystal engineering? Molecular crystals have interesting physical and chemical properties that are not associated with other categories of crystalline substances. These properties are connected to, and closely related to, their internal periodic structures. These internal structures are known as *crystal structures* and they are of outstanding importance in crystal engineering. So, there is a need to be able to design particular crystal structures, wherein molecules are assembled in particular ways. A particular crystal structure, in turn, has a particular property that is desired. Crystal engineering therefore consists of many different operations. These include the determination of crystal structures, the understanding or analysis of these and other known crystal structures, the use of this understanding in trying to design a crystal structure of a particular type including hitherto unknown structure types, the actual crystallization experiment, and finally the realization of a pre-desired crystal property. Clearly, many skills are involved in the art and science of crystal engineering. We will now trace the historical background of the several scientific streams of thought that have come together in this new subject.

1.1   X-ray Crystallography

Following the work of Max von Laué and the Braggs, father and son, and the visualization of a crystal as a periodic array that was capable of diffracting X-rays, the stage was set for the determination of crystal structures of pure chemical compounds. The earliest organic crystal structures to be determined were those of simple, symmetrical molecules like hexachlorobenzene and urotropin. Very early on, in 1921, W. H. Bragg commented that certain structural units like a benzene ring, having a definite size and form, might be retained with little or no change in going from one crystal structure to another. Bragg noted that the unit cell parameters of naphthalene and anthracene were related; two axial lengths were nearly the same while the third was 8.66 Å in naphthalene and 11.16 Å in

Zero, One, Two and Three-dimensional Molecules

One can use Kitaigorodskii's definition of a molecular crystal to obtain a useful geometrical concept of molecularity. A conventional molecule, say naphthalene, phenol or methane, is a zero-dimensional object. One finds limits, in all directions, to the molecular structure. Beyond a certain distance there are no more atoms in the molecule, in any direction. There are non-bonding boundaries in all directions. We can say accordingly that the molecularity of naphthalene or methane extends in all three dimensions. If we consider a crystal like $PdCl_2$, we find that there is an infinite linear bridged structure of Pd and Cl atoms. This structure constitutes a one-dimensional object. The molecularity of $PdCl_2$ extends into two dimensions but in the third dimension, which is along the length of the bridged structure, the molecule is extended infinitely. Let us next take graphite. Here the object is a two-dimensional molecule (sheet) and the molecularity extends only in one dimension, namely perpendicular to the sheet. Finally, extended solids like diamond and ZnS are giant three-dimensional molecular objects. Such ideas easily allow for coordination polymers (Chapter 7) to be defined as molecules and bring this large and interesting group of compounds well within the scope of crystal engineering.

anthracene. Accordingly, he concluded that the long axis of the molecules coincided with this third axis and further that the fused benzene ring dimension is approximately 2.50 Å. This is perhaps the earliest correlation between a crystal property and a molecular property, and is of relevance to modern crystal engineering because one of the fundamental questions of crystal engineering is *"Given the molecular structure of a compound, what is its crystal structure?"* The aims and goals of crystal engineering are well summarized in this question because one attempts, in this subject, to design crystal structures by using the molecule as a building block.

The first systematic answer to this fundamental question, which relates molecular and crystal structure, was given by J. M. Robertson, one of the most illustrious students of W. H. Bragg. Robertson, who worked for many years in the University of Glasgow, carried out a monumental study of the crystal structures of polynuclear aromatic hydrocarbons. This group of compounds includes not only simple compounds like naphthalene, anthracene and phenanthrene but larger and more complex molecules like perylene, coronene, ovalene and dibenzanthracene. These compounds are not only difficult to make and/or isolate but in the 1940s, when Robertson was carrying out many of his investigations, the crystal structure determinations of such compounds were of formidable difficulty. In 1951, Robertson concluded that these hydrocarbons could be classified into two groups. The first, in which the molecular area was small in comparison to the molecular thickness, is populated by hydrogen rich molecules like naphthalene and anthracene. The crystal structures in this group are characterized by short axes of around 5.0 Å; they contain molecules related by a *herringbone packing* of molecules inclined at about 40° to this short axis. The second group, in which the molecular area is large in comparison with the molecular thickness, is represented by carbon rich graphitic structures of coronene and ovalene. This classification is discussed further in Chapter 2. Robertson, in effect, carried out the first systematic experiment in crystal engineering. He identified a series of molecules, made them, determined their crystal structures and established a correlation between molecular structure and crystal structure. He was the first of a series of chemist-crystallographers, scientists who were adept in two

John Monteath Robertson (1900–1989).

rather different disciplines, and who did research in a topic where expertise in both disciplines was essential.

More about X-ray crystallography as a modern technique is given in Section 2.6.1.

1.2   Organic Solid State Chemistry

A coming together of crystallographic and chemical schools of thought, in the context of organic compounds, was seen more definitely during the period 1950–1980. This period saw the introduction of the terms *crystal engineering* and the allied term *crystal chemistry*.

In 1955, at a meeting of the American Physical Society in Mexico City, Ray Pepinsky first used the term *crystal engineering* when he stated that "crystallization of organic ions with metal-containing complex ions of suitable sizes, charges and solubilities results in structures with cells and symmetries determined chiefly by packing of complex ions. These cells and symmetries are to a good extent controllable: hence crystals with advantageous properties can be *engineered*." This definition encompasses the modern scope of crystal engineering as it is being practised 55 years later because it contains all three important elements of the subject: analysis, design and function. When this trio of attributes is taken together, our subject falls well within the scope of *engineering*.

The chemical slant of crystal engineering can surely be traced to the work of Gerhard M. J. Schmidt and his colleagues in the Weizmann Institute of Science during the period 1950–1970. Schmidt realized the power and significance of X-ray crystallography in organic chemistry. He laid the foundations of a new subject that he called *organic solid state chemistry*. Solution reactivity is largely a molecular property but solid state reactivity in crystals is characteristic of periodic assemblies of molecules. The link between structure and reactivity in the solid state structure was clearly established by Schmidt for a series of photodimerizable alkenes. Taking substituted *trans*-cinnamic acids as examples he studied their 2+2 photocycloaddition reactions to cyclobutanes. He found that for these compounds, the crystalline matrix provides an extraordinary spatial control on the initiation and progress of the solid state reaction. For a start, the cinnamic acids do not undergo dimerization in solution, and even if they do, there are many products and the overall conversion to dimer is

Gerhard Schmidt (1919–1971) is widely credited with the introduction of the term *crystal engineering* into the chemical literature in 1971. However, it is clear that the term was first used by Ray Pepinsky in 1955 in more or less the same context that Schmidt did, and largely in the sense that we use today. Schmidt began his work on the topochemistry of crystalline alkenes in the early 1950s. He only began systematic publication of his results in 1964 in a long series of monumental back to back papers in the *Journal of the Chemical Society*. It is also known that Schmidt and Pepinsky knew each other. Pepinsky's reference to the term *crystal engineering* is traced to just one occurrence in 1955. Schmidt published a voluminous amount of material on the subject. Did they discuss these ideas together?

low. In the solid state, however, the conversion efficiencies are high and the regiospecificities of the obtained products are clear and decisive. When the crystal contains nearest neighbour molecules in which the reacting double bonds are related by a crystal inversion centre, the cyclobutane that is obtained has inversion symmetry. When the nearest neighbour molecules are related by translation symmetry in the crystal, the product has mirror-symmetry.

Let us illustrate these principles further. Substituted cinnamic acids crystallize in one of three structural types α, β and γ and while the α and β forms react in a 2+2 manner to give cyclobutanes when irradiated in the solid state, the γ forms are photostable. Schmidt assumed that reactivity in the solid state takes place with a minimum of atomic and molecular movement, the *topochemical principle*. Accordingly, the formation of mirror symmetry truxinic acids from the β-cinnamic acids and that of the inversion symmetry truxillic acids from the α-acids becomes intuitively obvious, as is the photostability of the γ-acids.

Because this on-off model for predicting the outcome of a solid state reaction was so successful, attention became directed to the control of the crystal structures themselves. This proved to be much more difficult to do. Simple substitutions on a molecular skeleton usually caused deep-seated changes in the crystal structure. Therefore the concept of crystal engineering came into being. The term was employed by Schmidt to denote rules that could be used to generalize and predict the packing of molecules within crystals. In 1971, just before his death, he stated in a now well known paper in the journal *Pure and Applied Chemistry* that "the systematic development of our subject will be difficult if not impossible until we understand the intermolecular forces responsible for the stability of the crystalline lattice...any rational development of the physics and chemistry of the solid state must be based upon a theory of molecular packing; since the molecules studied are complex, the theory will be empirical for some time yet. Rules are now becoming available in what I would regard as phase three, *the phase of crystal engineering*."

In the 15 or so years that followed the death of Gerhard Schmidt, chemical crystallographers made some efforts aimed at the design of reactive solids. Topochemical 2+2 reactions were designed by scientists in the Weizmann Institute, notably Mendel Cohen, Meir Lahav and Leslie Leiserowitz. A group in

Possibly the first "rule" of crystal engineering was that developed by G. M. J. Schmidt and B. S. Green who found that aromatic molecules that are dichlorosubstituted have a marked tendency to adopt a crystal structure with a short axis of around 4 Å. This is equivalent to the β-structure of cinnamic acid. This empirical "chloro rule" was used to make alkenes that would undergo photodimerization to give mirror symmetry cyclobutane derivatives.

Cambridge, headed by John M. Thomas developed a series of cyclic ketones, namely the 2-benzyl-5-benzylidenecyclopentanones, which showed an interdependence of crystal structure and solid state reactivity. Gerhard Wegner, in Freiburg, showed that the polymerization of crystalline diacetylenes is governed by topochemical rules. Masaki Hasegawa and others showed that the solid state polymerization of 2,5-distyrylpyrazine is very similar to that of the diacetylenes in that there is a close relationship between the crystal structure of the starting monomer and the likelihood of a solid state reaction taking place.

Another type of solid state chemical reactivity was exemplified by the reactions of organic crystals with gases. In the University of Illinois, David Y. Curtin and Iain C. Paul discovered a striking anisotropy in the reactions of ammonia with crystalline benzoic acids. Reaction was fastest on those faces where carboxylic groups emerged and slowest on those where the hydrocarbon fragments were situated. Solid-gas reactions are important in some industrial processes but are difficult to predict, organize and control. Once again, however, a connection was established between internal crystal structure and molecular packing, which are microscopic properties, and bulk chemical reactivity which is a macroscopic property. The fact that such connections do exist makes it important to achieve some measure of control over crystal packing. How does one predict the crystal structure of an organic compound? We come back to the basic question: Given the molecular structure of an organic compound, what is its crystal structure?

Chemist crystallographers in the 1970s and early 1980s were, on the whole, restricting their attention to organic reactions in the solid state. These systems were hard to study. The routine determination of organic crystal structures was not yet quite a reality and methods of studying molecular solid state reactions were not similar to the methods employed in the solid state chemistry of extended inorganic solids. In consequence, this period saw something of a hiatus.

1.3   The Crystal as a Supramolecular Entity

A distinct thought stream, and one which was to ultimately bring crystal engineering into the chemical

The synthetic chemist is likely to have many opportunities to encounter unusual phenomena by accident during everyday chemical work with crystalline solids and without the proper background will not be prepared to recognize and take advantage of such chance discoveries. There is the further misfortune that those workers in areas where the most dramatic applications...have occurred are, in general, uncomfortable when dealing with structures of complex organic molecules; the result is a serious lack of communication between groups whose interaction should be mutually beneficial.

D. Y. Curtin and I. C. Paul, *Chem. Rev.*, 1981.

Aristotle is widely credited with introducing holistic thinking into science, although this type of thought was also widely prevalent in ancient India and China, forming the basis of many current Eastern systems of philosophy. Aristotle believed in the predominance of the species rather than the individual members of the species. In the context of supramolecular chemistry, for example, the network is more important than the molecules that constitute it. The whole is greater than the sum of the parts. Holism stands in direct contrast to reductionism which stresses the importance of the individual and the uniqueness of everything. This school of thought was originally propagated by Charles Darwin. According to reductionist thinking a complex phenomenon can be fully understood by dissecting it down to simpler components and understanding their individual behavior. According to reductionist thinking, chemistry is reducible to physics, and physics in turn is reducible to mathematics. Supramolecular chemistry is perhaps the first important thought stream in the history of chemistry that is blatantly non-reductionist in its approach.

mainstream was the theme of *supramolecular chemistry*. Chemistry, especially organic chemistry, was a wholly molecule centric subject till around 50 years ago. The molecule was considered to be the ultimate delimiter of all significant properties of a substance. By looking at a molecular structure, the chemist was able to conceive of all the useful and important attributes of a chemical compound. The molecular paradigm of organic chemistry was the inspiration for synthetic chemistry. The philosophy of organic synthesis and the way it grew over the last century is described in Chapter 3. The fact that there are some fundamental properties of a compound, like its melting point, that depend on the structure of groups or assemblies of molecules was either ignored or not considered carefully enough.

The development of supramolecular chemistry is outlined in Section 3.2. In this introductory chapter, it will be sufficient to mention that the first modern stirrings of supramolecular thought occurred in the description of an organic crystal structure as a network by H. M. Powell who, in 1948, discussed the hydroquinone clathrates in this context. A network description of a structure is a holistic description that de-emphasizes the constituents of the network, namely the molecules. Topology and form take precedence over content. A major premise in supramolecular chemistry is that the whole is greater than the sum of its parts. An assembly or collection of molecules has a structure and properties that are more than the sum total of the properties of the individual molecules.

The supramolecular concept successfully draws together organic, inorganic and organometallic chemistry in the context of crystal engineering. This conceptual coming together of disciplines that were traditionally considered to be distinct occurs through the visualization of crystal structures as networks. An organic crystal structure, in other words, a molecular crystal structure, was usually viewed as a collection of molecules and not as a network. Inorganic infinite solids, on the other hand, are more often than not viewed as networks that are made up of atoms or ions (Section 1.4.3). With supramolecular chemistry, an organic crystal may be depicted as a network. One can also look at an inorganic solid as a "molecular crystal". Let us discuss these interpretations further.

In 1988, Otto Ermer, in the University of Cologne, provided a beautiful and, at the time, unique example of a three-dimensional hydrogen bonded network in his description of the hydrogen bonded crystal structure of adamantane-1,3,5,7-tetracarboxylic acid. The structural formula of this acid is shown in the inset. The formation of hydrogen bonds at each of the four tetrahedrally disposed carboxyl groups to other such groups on adjacent molecules leads to a diamondoid lattice (see also Section 3.3.2). This structure is topologically equivalent to the diamond lattice with the C–C bonds replaced by O–H$\cdots$O hydrogen bonded dimers. This structure deviates considerably from close packing. The open diamondoid network creates a large empty hollow within, but this problem is solved by the *interpenetration* of no less than five such diamondoid networks. Notably, there is no hydrogen bonding between networks. The appearance of this paper in the late 1980s was significant. Clearly the scientific community was ready for supramolecular thinking. The Nobel Prize had just been awarded to Charles J. Pedersen, Donald J. Cram and Jean-Marie Lehn in 1987 for supramolecular chemistry. Chemists were finally willing to consider holistic approaches. The depiction of an organic crystal structure as a network is one such non-reductionist approach.

In the end, *all* organic crystal structures can be viewed at as networks. The chemist emphasizes the linkages between molecules rather than the molecular structure itself. For example, the orthorhombic crystal structure of benzene approximates a face centred cubic packing (FCC), and this is not surprising, given the hydrocarbon nature of the molecule and its discoid shape. The more spherical urotropin forms an accurately body centered cubic (BCC) crystal and there are not many conceptual differences between such structures and those of the metallic elements. All organic crystal structures may be considered as networks with the molecules being the nodes and the non-bonded intermolecular interactions being the node connections (Section 1.4.3). The dissection of an organic crystal structure into one, two or three-dimensional networks would become very important in crystal engineering in the years to come because such an analysis reveals similarities between crystal structures of molecules with widely different molecular structures. This can occur because the node connections in a network can be formed from a wide variety of intermolecular interactions.

Traditionally, organic and inorganic crystal chemistry have proceeded along different lines of thought even though propositions such as the need to fill space most economically and the conflict between close packing and directionality are of universal applicability. In some instances, it is quite useful to seek similarities between organic and inorganic crystal structures. Such analysis is extremely important in the study of coordination polymers (Chapter 7). These similarities are topological in nature with equivalences being found between intermolecular interactions in organic solids and ionic and covalent bonds in extended inorganic solids. Just as an organic structure can be viewed as a network with nodes (molecules) and node connections (interactions), an inorganic crystal structure can be looked upon as a giant molecule using the concepts of molecularity as outlined earlier in this chapter. Diamond and cubic ZnS are indeed giant molecules and there is no difficulty in such visualization. For ionic structures like NaCl and CsCl, a "molecular" visualization is nominally possible if the ions are likened to "atoms" and the ionic bonds to "atom connections". Considering also that covalent bonds and intermolecular interactions form a continuum and that the strongest interactions are stronger than the weakest bonds, it is possible to find an inorganic topological counterpart for many organic structures. The analogy between adamantane-1,3,5,7-tetracarboxylic acid and diamond is described here. Some others are mentioned in the set of problems at the end of this chapter.

The fundamental laws necessary for the mathematical treatment of a large part of physics and the whole of chemistry are thus completely known, and the difficulty lies only in the fact that application of these laws leads to equations that are too complex to be solved.

This famous statement, which is dismissive of chemistry as a whole, was made by the famous physicist P. A. M. Dirac and could be termed the reductionist's high noon.

Wilhelm Ostwald (1853–1932), one of the founding fathers of physical chemistry.

1.4   Modern Crystal Engineering

A new scientific subject generally emerges when there is a critical mass of thought coupled with advances in instrumentation. The emergence of a new subject is, however, not fully predictable and can be influenced in part by random events. The appearance of a new area is usually triggered by a few critical papers, a small number of individuals, and some opinion-making conferences. In any event, the scientific climate should be such that there is a willingness to accept new ideas. The arrival of modern crystal engineering in the late 1980s and early 1990s arose from a combination of expected and unexpected reasons. Let us first examine how chemistry has been organized as a subject in the past.

1.4.1   *Horizontal and Vertical Divisions of Chemistry*

Chemistry was perceived as a distinct subject following John Dalton's atomic theory and its history over the past 200 years represents a consolidation of reductionist thinking, which brought about a division into organic, inorganic and physical streams. Chemistry in the 19th and 20th century was seen as being more or less reducible into physics and finally mathematics. Depending on the importance of quantitative ideas as manifested in the ease of reducibility, these three streams of chemistry came into being. Organic chemistry, the most qualitative and irreducible of the streams, developed as a separate subject after Friedrich Wöhler's urea synthesis in 1828 and was strengthened by the work of Justus Liebig, August Kekulé and others. Physical chemistry became clearly demarcated with Wilhelm Ostwald's contributions in the later part of the 19th century, and underwent something of a metamorphosis after Linus Pauling's work in the 1930s on the chemical bond. This set the stage for the appearance of quantum chemistry, which was always considered to lie within physical chemistry. Inorganic chemistry, the part of chemistry that did not "go away" became greatly focused after Alfred Werner's work on coordination compounds (Chapter 7). In the decades that followed, considerable overlap was found between physical and inorganic chemistry. Solid state chemistry (of inorganic compounds) lies in this intersection. Some chemists consider analytical chemistry to rank along with organic chemistry, inorganic chemistry and physical chemistry as the fourth fundamental pillar of the subject. These are the vertical divisions of chemistry. These divisions do not carry the same significance today, in a research sense, that they did say 50 years ago but they are still very convenient demarcations in a pedagogic sense.

In the last part of the 20th century, several other branches of chemistry came into prominence. These new subjects resisted an easy docketing into one of the three vertically divisible streams of chemistry. Therefore they may be called horizontal divisions. They include medicinal, environmental, theoretical, computational, solid state, pharmaceutical, materials and supramolecular chemistry. These subjects are so interdisciplinary and eclectic that it is scarcely possible, and not even necessary, to try and pigeon hole them into the conventional organic-inorganic-physical classification. Crystal engineering is one of these new horizontal branches of chemistry. It takes ideas from organic, inorganic, physical, computational, supramolecular and solid state chemistry, and surely from X-ray crystallography. Some of these thought streams have been described earlier in this chapter. Others will be discussed in subsequent chapters. The appearance of modern crystal engineering on the scientific scene owed in no small measure to the award of the 1987 Nobel Prize to work in supramolecular chemistry. Crystal engineering is noteworthy in that it represents, like supramolecular chemistry, a departure from reductionist thinking. In this new non-reductionist chemistry, a case is made for the emergence of a new set of properties from the interplay of macrosystems that are not related directly to their component atoms and molecules. The idea of emergence was linked to complex pathways and was used to explain the evolution of complex self-organizing systems, like molecular crystals. By the late-1980s, the time was ripe for the launching of new areas within the chemical sciences. Crystal engineering was one of these new areas.

Not included in Section 1.4.1 is a description of the areas of overlap between chemistry and biology. Biochemistry emerged from chemistry around 1900 with the work of Emil Fischer (1852–1919) but it is not even considered a part of chemistry today. Systems biology is more closely related to modern day supramolecular chemistry and much interdisciplinary work is carried out under the umbrella of chemical biology. Curiously, Fischer is also credited with the first mention of supramolecular ideas in his lock-and-key model for the working of enzymes (Section 3.2). It is surely not a coincidence that the same chemist initiated thinking in both biochemistry and supramolecular chemistry.

1.4.2 *Organic Crystal Engineering*

Several factors were responsible for the evolution of modern crystal engineering in the late 1980s. Notable among them was the fact that it became easy, around then to determine the crystal structures of small organic molecules. This was due to advances both in

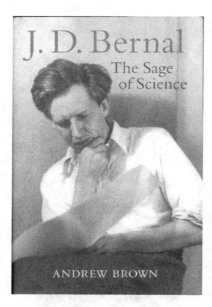

Bernal suggested in the 1930s that collections of crystallographic data would be of value.

instrumentation (computer controlled single crystal diffractometers) and computation (Direct Methods for structure solution and programs such as SHELX written by George Sheldrick). The second was the emergence of the Cambridge Structural Database as a research tool of importance. A paper by Olga Kennard, Frank Allen and Robin Taylor in *Accounts of Chemical Research* in 1983 was to have far-reaching importance. The CSD could be used for *data mining* of crystal structure information. It became possible to analyze and identify recurring patterns of molecules in organic crystals. A recurring pattern is more likely to repeat itself in a new crystal structure, and the ground rules of crystal structure prediction thus laid. There was also the increasing perception that if crystal engineering was to live up to its promise of becoming a new and interesting subject, it should outgrow the then-existing emphasis on solid state reactivity. Other properties were of interest too. Fundamental questions were being asked as to the underlying causes for the adoption of crystal structures. Organic crystal structures were not even categorized or classified into types, as had been done for inorganic solids. There were stirrings of new ideas with respect to intermolecular interactions, notably the weaker varieties of hydrogen bonds. A seminal paper in *Acta Crystallographica* by Leslie Leiserowitz, in 1976, on the crystal chemistry of carboxylic acids was an early hint of these new ideas. Could weak hydrogen bonds possibly be important in organic crystal structures?

There is general consensus that many present-day motivations of crystal engineering originated from a book that appeared in 1989, authored by G. R. Desiraju, entitled *Crystal Engineering. The Design of Organic Solids*. Notable in this book are the following points of departure from the then-existing paradigms: (i) Crystal engineering is a subject of wide-ranging applications and its scope extends much beyond the topochemical photoreactivity of organic solids; (ii) Organic crystal structures are predominantly governed by Kitaigorodskii's close packing principles, which invoke geometrical arguments, but the minor deviations from close packing, which owe to chemical factors, are of the greatest importance because they lead to the formation of crystal structures that can be *engineered* in a systematic manner. Directionality, such as it exists in organic crystals, is the handle that permits crystal design; (iii) The collection of machine retrievable data on 70,000 crystal structures that existed at that time (the number is now greater than 525,000) is sufficient to establish trends in crystal packing. Pattern recognition is one of the first steps in crystal engineering strategy; (iv) Directional interactions are anisotropic in nature and they cause deviations from close packing. These interactions need to be manipulated in the crystal design exercise. The most important of these directional interactions is hydrogen bonding, which exists in strong and weak varieties; (v) Other directional interactions are based on halogen and sulfur atom non-bonded contacts in crystals; (vi) The design of crystals that lack an inversion centre was identified as a systematic endeavour in crystal engineering; (vii) Multi-component molecular crystals, the study of which is seen as an activity of the greatest importance in modern crystal engineering, were identified as a distinct sub-group of structures which might be amenable to distinct methods of design; (viii) Polymorphism, the existence of multiple crystal forms for the same molecule, was labeled as a Nemesis of crystal engineering, thwarting the plans of the crystal engineer.

Some of these ideas will be elaborated in this textbook. In Chapter 2, we will discuss close packing and directional intermolecular interactions. In Chapter 5, we will describe polymorphism in all its ramifications and understand why this subject, while still being a Nemesis in terms of its difficulty, has come to be of the greatest practical significance. In Chapter 6, we will examine multi-component crystals and try to understand if they are similar to single component crystals.

A number of important papers and studies quickly followed. A set of three papers co-authored by Desiraju and Angelo Gavezzotti classified the crystal structures of polynuclear aromatic hydrocarbons and identified the molecule → crystal paradigm as a primary concern of crystal engineering. Influential papers by Margaret Etter focussed on hydrogen bonding as a design tool. She termed hydrogen bonds as being both strong and directional and described a method of classifying and labelling hydrogen bond networks; this is detailed in Section 2.6.2.1. In hindsight, it is interesting to observe that both these sets of studies reinterpreted the early works of Robertson: the Desiraju-Gavezzotti papers expanded and quantified his studies of hydrocarbons while Etter did practically the same thing with her work on hydrogen bonding. Such reinterpretation is almost mandatory in an interpretative science, which is what crystal engineering is all about.

Nemesis, the Greek goddess of retribution, symbolizes the spirit of divine vengeance against those who submit to arrogance. There is something of the inevitable in this kind of divine wrath. Implied almost is that there is a modicum of good and bad for all of us. Nemesis is the distributor of fortune in due proportion. Undeserved good fortune would attract her attentions and would not go unpunished. In science too, it is interesting to wonder if such themes actually prevail. The subject of polymorphism in crystal engineering is full of anecdotes (Chapter 5) in which punishment, in the form of a disappearing polymorph or a contentious law suit, has often thwarted the moves of the rich and powerful. Perhaps scientists too need to pay heed to some of the lessons learned from Greek mythology.

Participants in the crystallography school in Erice in 1985.

Gradually the community of crystal engineers began to grow. An important tutorial meeting in Erice, Sicily, in 1985, chaired by Aldo Domenicano, Istvan Hargittai and Peter Murry-Rust might already have sowed the seeds for some of the developments in crystal engineering in the late 1980s. Many scientists, who are currently prominent in the field, attended this meeting during their formative years. This meeting highlighted the fact that X-ray crystallography of small organic molecules was becoming easy, accurate and fast. In the 1990 Congress of the International Union of Crystallography in Bordeaux, there was free discussion on hydrogen bonds as agents of molecular association in crystals. The emphasis was shifting surely from structures of individual molecules to surveys of intermolecular packing and interactions, *crystal structures*. Along these lines, there were papers by Michael Zaworotko on hydrogen bonding and diamond networks, and by Dario Braga on organometallic cluster compounds in the early 1990s. Others too took up the issue of molecular aggregation in more general ways. A time line of crystal engineering over the last century is given at the end of this chapter and will give the reader a glimpse of some important developments.

Desiraju's 1989 book deals with the crystal engineering of organic solids. In a parallel development that was influenced by the idea of a crystal structure as a network, crystal engineering of metal-organic crystal structures was mooted.

1.4.3   *Metal-Organic Crystal Engineering*

We have already seen that inorganic extended solids are typically described as networks (Section 1.3). This depiction was made popular by A. F. Wells, whose monumental book is now a classic. Wells defined crystal structures in terms of their topologies and reduced them to a collection of nodes (points) of particular geometry (tetrahedral, trigonal, planar) that are connected to each other with node connections (bonds, interactions). Any network consists of nodes and node connections. The resulting structures can also be calculated mathematically and can be infinite in one, two or three dimensions depending on the chemical nature of the nodes and node connections.

Richard Robson extended the work of A. F. Wells on inorganic extended solids to coordination compounds. By linking polycoordinate metal ions with polydentate ligands, it is possible to build up a polymeric structure. Therefore these compounds were called *coordination polymers*. They are described in detail in Chapter 7. Here, it will suffice to state that a network depiction for coordination polymers is intuitive. The metal centers act as nodes and the ligands behave as node connections. The first compound Robson made in 1989, $[Cu^I\{C(C_6H_4CN)_4\}]^+$, has a diamond network because it is constructed from a tetrahedral metal center (Cu^I) and a tetradentate ligand. The intention, fully realized, in this crystal engineering exercise was that the metal ion would bind to a tetrahedrally disposed set of four nitrile donors from separate ligand molecules, each of which in turn binds to four metal ions, extending the structure infinitely in three dimensions. The cavities created by this network are rather large and are filled with solvent and counterions.

We will describe the crystal structure of $Zn(CN)_2$ in Chapter 7. At this stage, it is sufficient to state that this solid too has a diamond network structure but both C and N atoms in the CN ligand can act as donors, and therefore each Zn^{II} ion is coordinated to

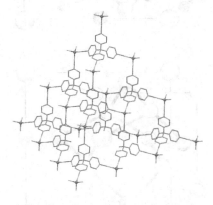

Tetrahedral network in $[Cu^I\{C(C_6H_4CN)_4\}]^+$

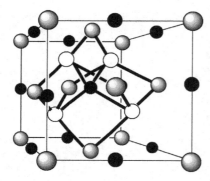

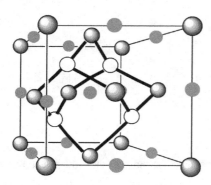

In 1994, G. R. Desiraju described an organic multi-component system in terms of networks. 1,3,5,7-Tetrabromoadamantane (large grey shaded) forms a 1:2 diamond network crystal with urotropin. There are two crystallographic sites for urotropin (black and unshaded). One of them (black) is loosely bound and can be substituted with CBr_4 (light grey) a molecule of the same size. These experiments are very similar to Robson's 1990 studies on $Zn(CN)_2$ in that both are based on retrosynthesis, a concept that was already well established in synthetic organic chemistry, but still had to become formalized in crystal engineering. Such formalization occurred with the publication of Desiraju's 1995 paper on supramolecular synthons (Chapter 3).

two C-atoms and two N-atoms. This network is doubly interpenetrated, in other words two such networks catenate within each other with no formal link between them, similar to adamantane-1,3,5,7-tetracarboxylic acid which has a fivefold interpenetrated structure (Section 1.3) Noting that the affinities of Cu^I and Zn^{II} cations to C-atom and N-atom donors are different (Cu^I binds preferentially to C-donors), Robson conjectured correctly that if half the Zn^{II} sites in $Zn(CN)_2$ were replaced with Cu^I, an ordered structure would be obtained in which the Zn and Cu sites alternate strictly (the Zn coordinating to the N-donors and the Cu to the C-donors). He further extrapolated correctly that if a counterion of the correct size were chosen (he picked Me_4N^+), it would fit exactly into the cavities of the tetrahedral network so that interpenetration, as in $Zn(CN)_2$ would be avoided. Robson's work was followed by that of Omar Yaghi who, in 1995, first proposed that guest molecules could be absorbed and removed from the open pores of coordination compounds. These compounds were called metal-organic framework compounds or MOFs.

The fields of organic crystal engineering and coordination polymers were viewed, in the past, as being somewhat distinct but there is now a growing realization that they are conceptually very similar and also that they have borrowed from the same chemical thought streams. Retrosynthetic methods (Chapter 3), for example, are an integral part of crystal design strategies in both branches of the subject. In the modern day context, both these areas are considered to lie within the scope of crystal engineering.

1.4.4 *Properties of Crystals*

Crystal engineering is all about making crystals that have specific functions. It is about the systematics of crystal construction. In his legendary talk in 1959 entitled "There's plenty of room at the bottom", Richard Feynman alluded to the problem of synthesis as something a physicist might want to attempt. He asked "Is there a *physical* way to synthesize any chemical substance?" He continued and said that, in principle, no fundamental law of physics was violated if a molecule were assembled systematically atom by atom. Similarly, we can say that no law would be violated if a crystal is put together systematically molecule by molecule. In the end, this is what crystal engineering is all about. Ultimately we want to make crystals that have a particular property, and we want to get control over the ways in which molecules assemble.

Crystal properties are both chemical and physical. Many properties are directly related to the crystal structure. Chemical properties lead to a change in chemical composition of the crystal, which may be reversible or irreversible. Molecular crystals with particular chemical properties (reactivity, tautomerism, changes in color) find applications as sensors, devices, light sensitive materials, and in catalysis. Physical properties (gas inclusion, electrical, magnetic, optical, solubility) are more easily tailored and the majority of applications are in this category. Nonlinear optical responses of some organic solids that lack a crystallographic center of symmetry exceed that found in presently used inorganic substances. These crystals are used as frequency doublers and waveguides. Noncentrosymmetric crystals also find applications as ferroelectrics, piezoelectrics, pyroelectrics and as triboluminescent materials. Solid form control is important in industries connected with dyestuffs, pharmaceuticals and explosives. The importance of obtaining different solid forms of a drug molecule (polymorph, solvate, co-crystal) can be of huge commercial interest. A major application of molecular crystals today is in the engineering of open metal-organic framework structures (Chapter 7). These compounds have been compared to zeolites because they have framework structures with large spaces enclosed within. Compared with zeolites, however, the dimensions of the empty spaces in coordination polymers are large and the internal surface area is concomitantly large. Accordingly, these materials are able to absorb large quantities of gases. They can also be engineered so that they can selectively absorb gases from gas mixtures. Such applications can be immensely useful in separation and catalysis.

Richard Feynman (1918–1988).

1.5 Summary

This book is written so that it can be read all at once, or chapter by chapter. It is meant for senior undergraduates and beginning graduate students. In Chapter 2, we will discuss intermolecular interactions, the glue that holds molecules together in crystals. In Chapter 3 is outlined the main strategy for retrosynthesis of an

Reason is to the imagination as the instrument to the agent, as the body to the spirit, as the shadow to the substance.

Shelley

Claude Berthollet (1748–1822).

Johann Wolfgang von Goethe (1749–1832).

organic crystal. Crystal engineering is described as supramolecular synthesis. Chapter 4 provides a summary of the main issues connected with crystallization. Understanding the mechanism of crystallization is considered to be one of the holy grails of crystal engineering. In Chapter 5 is provided an account of polymorphs, or different crystal forms of the same compound. Is polymorphism the Nemesis of crystal engineering? Chapter 6 contains a description of multi-component molecular crystals or co-crystals. Are these really different from single component crystals? Chapter 7 gives an account of coordination polymers or metal-organic framework compounds and will show how the study of these crystalline substances properly comes within the scope of crystal engineering. A glossary of useful terms is given at the end of the book. Nomenclature issues have always been lively in crystal engineering, and it is hoped that the annotations given in the glossary are both noncontroversial and useful. Also provided, at the end of each chapter, is a list of references for further study. This is a textbook and not a research book. Therefore, we have avoided mentioning references to the literature in the main body of the text. Representative problems are also given in each chapter.

Crystal engineering has been described as the designed synthesis of functional crystalline matter. In the end, however, crystal engineering is a form of synthesis and, in this respect, it is firmly a part of the chemical enterprise. Chemistry is the only important subject in which the researcher is free to make the object of his study. It was said by Berthollet that "La chime crée son objet". All synthesis is a combination of art and science, the substance and the shadow, and crystal engineering is no exception. Why do we make a crystal? Because it is there in the mind, waiting to be made; because the play of symmetry in the crystal is aesthetically pleasing; because it is difficult to make; because it is useful; because it is useless. Synthesis needs no justification. It is an integral part of chemistry. Some would say that synthesis is chemistry itself. Crystallography has its own fatal attractions. In the words of Goethe, who said in 1822, "Crystallography, considered as a science, gives rise to quite specific viewpoints. It is not productive, and it exists only for itself, having no consequences, especially now that one has encountered so many isomorphic substances. As it is really nowhere useful, it has developed largely into itself. But it does provide the intellect a certain limited satisfaction, and its details are so diverse that one can describe it as inexhaustible and that is why it binds even first rate people so firmly and for so long." The sciences of chemistry and crystallography have had a deep and symbiotic relationship for a great many years. Crystal engineering is one the latest, and most exciting, examples of this symbiosis.

The first scientific meeting devoted exclusively to crystal engineering was organized by M. J. Zaworotko and K. R. Seddon in Digby, Nova Scotia, Canada in 1996.

Crystal Engineering Time Line

1921	W. H. Bragg relates the crystal structures of naphthalene and anthracene.
1935	J. D. Bernal advocates the study of groups of crystal structures. The forerunner of crystallographic databases.
1948	H. M. Powell describes the β-hydroquinone structure as a network.
1951	J. M. Robertson's paper on the crystal structures of polynuclear aromatic hydrocarbons.
1954	Crystal structure of the Hofmann complex by J. H. Rayner and H. M. Powell.
1955	R. Pepinsky uses the term *crystal engineering*, the first time anyone does so.
1959	R. Feynman's lecture on building a structure from bottom up. There's room at the bottom.
1964	G. M. J. Schmidt publishes a series of papers in *Journal of the Chemical Society* on the solid state chemistry of cinnamic acids and other alkenes. The topochemical principle.

1971	Schmidt formally introduces the term *crystal engineering* in his paper in *Pure and Applied Chemistry.*
1970–1985	Development of organic solid state chemistry in the Weizmann Institute (M. D. Cohen, M. Lahav, L. Leiserowitz), University of Illinois (D. Y. Curtin, I. C. Paul), University of Freiburg (G. Wegner) and University of Cambridge (J. M. Thomas).
1977	Crystal structure determination of Prussian Blue by A. Ludi *et al.* published in *Inorganic Chemistry.*
1988	First analysis of an interpenetrated organic solid. O. Ermer's paper in *Journal of the American Chemical Society* on adamantane-1,3,5,7-tetracarboxylic acid.
1989	Robertson revisited. Paper by G. R. Desiraju and A. Gavezzotti entitled *From molecular to crystal structure* in *Chemical Communications.*
1989	G. R. Desiraju's book *Crystal Engineering. The Design of Organic Solids* is published signifying the beginning of modern organic crystal engineering. This is still the only single author book on the subject.
1990	R. Robson's paper in *Journal of the American Chemical Society* on interpenetrated transition metal coordination compounds. The field of coordination polymers is born.
1990	M. C. Etter identifies the hydrogen bond as an important design element in crystal construction in her review in *Accounts of Chemical Research.*
1991	J. D. Dunitz terms the crystal as a *supramolecule par excellence* in his paper in *Pure and Applied Chemistry.*
1991	J. D. Wuest introduces the concept of molecular tectonics. This term was originally introduced by N. N. Greenwood in 1984 in the context of polyhedral boron clusters.
1993	Trial of Glaxo's action against Novopharm, a Canada-based generic company, for infringement of one of Glaxo's patents covering the block-buster anti-ulcer drug Zantac. The importance of polymorphism in crystal engineering becomes apparent.
1995	The relationship between crystal engineering and organic synthesis is highlighted in a review by G. R. Desiraju in *Angewandte Chemie.* Supramolecular retrosynthesis is introduced with the term *supramolecular synthon.*
1995	O. M. Yaghi describes selective binding and removal of guest molecules in a porous metal organic framework structure. Functional coordination polymers are launched.
1996	The first keynote lectures on crystal engineering in a Congress of the International Union of Crystallography are delivered by A. Gavezzotti and G. R. Desiraju in Seattle.
1996	The first scientific meeting devoted entirely to crystal engineering is organized by M. J. Zaworotko and K. R. Seddon in Nova Scotia.
1998	Crystal engineering becomes a microsymposium topic for the first time in a Congress of the International Union of Crystallography in Glasgow.
1999	The Royal Society of Chemistry launches the journal *CrystEngComm* with D. Braga as the chairman of the editorial board.
2001	The American Chemical Society launches the journal *Crystal Growth & Design* with R. D. Rogers as the editor-in-chief.
2004	M. J. Zaworotko and Ö. Almarsson describe the importance of pharmaceutical co-crystals in a review in *Chemical Communications.*

2004	S. Kitagawa describes third generation coordination polymers which the cavities in the metal-organic framework structure are responsive to guest shape.
2010	The first Gordon Research Conference in Crystal Engineering is held in New Hampshire.
2011	The first textbook devoted exclusively to crystal engineering is published and is co-authored by G. R. Desiraju, J. J. Vittal and A. Ramanan.

1.6 Further Reading

Books

J. M. Robertson, *Organic Crystals and Molecules*, 1953.

A. I. Kitaigorodskii, *Molecular Crystals and Molecules*, 1971.

J. D. Dunitz, *X-Ray Analysis and the Structure of Organic Molecules*, 1981.

G. R. Desiraju, *Crystal Engineering. The Design of Organic Solids*, 1989.

J. D. Wright, *Molecular Crystals*, 1995.

J.-M. Lehn, *Supramolecular Chemistry. Concepts and Perspectives*, 1995.

G. A. Jeffrey, *An Introduction to Hydrogen Bonding*, 1997.

J. W. Steed and J. L. Atwood, *Supramolecular Chemistry*, 2000.

J. Bernstein, *Polymorphism in Molecular Crystals*, 2002.

A. Gavezzotti, *Molecular Aggregation: Structure Analysis and Molecular Simulation of Crystals and Liquids*, 2007.

K. C. Nicolaou and T. Montagnon, *Molecules that changed the World*, 2008.

L.-L. Ooi, *Principles of X-ray Crystallography*, 2010.

Papers

D. E. Palin and H. M. Powell, The structure of molecular compounds 5. The clathrate compound of quinol and methanol, *J. Chem. Soc.*, 571–574, 1948.

M. D. Cohen and G. M. J. Schmidt, Topochemistry 1. Survey. *J. Chem. Soc.*, 1996–2000, 1964, and the succeeding papers.

G. M. J. Schmidt, Photodimerization in the solid state, *Pure Appl. Chem.*, 27, 647–678, 1971.

L. Leiserowitz, Molecular packing modes. Carboxylic acids. *Acta Crystallogr., Sect. B*, 32, 775–802, 1976.

F. H. Allen, O. Kennard and R. Taylor, Systematic analysis of structural data as a research technique in organic chemistry, *Acc. Chem. Res.*, 16, 146–153, 1983.

O. Ermer Fivefold diamond structure of adamantane-1,3,5,7-tetracarboxylic acid, *J. Am. Chem. Soc.*, 110, 3747–3754, 1988.

G. R. Desiraju and A. Gavezzotti, From molecular to crystal structure. Polynuclear aromatic hydrocarbons, *J. Chem. Soc. Chem. Commun.*, 621–623, 1989.

B. F. Hoskins and R. Robson, Infinite polymeric frameworks consisting of three dimensionally linked rod-like segments, *J. Am. Chem. Soc.*, 111, 5962–5964, 1989.

M. C. Etter, Encoding and decoding hydrogen bond patterns of organic compounds, *Acc. Chem. Res.*, 23, 120–126, 1990.

J. D. Dunitz, Phase transitions in molecular crystals from a chemical viewpoint, *Pure Appl. Chem.*, 63, 177–185, 1991.

S. Subramanian and M. J. Zaworotko, Exploitation of the hydrogen-bond — recent developments in the context of crystal engineering, *Coord. Chem. Rev.*, 137, 357–401, 1994.

M. J. Zaworotko, Crystal engineering of diamondoid networks, *Chem. Soc. Rev.*, 23, 283–288, 1994.

D. Braga and F. Grepioni, From molecule to molecular aggregation — clusters and crystals of clusters, *Acc. Chem. Res.*, 27, 51–56, 1994.

1.7   Problems

You may need access to the internet to attempt many of these problems. You may also need special (not free) software in some cases.

1. Refer to the definition of crystal engineering at the beginning of this chapter and verify that this definition addresses all three aspects of the subject: intermolecular interactions, crystal design and properties of molecular crystals.
2. What are the crystal structures of ZnS, CsCl, NbO and PtS? Topologically speaking, how do these compare with the well known structure of NaCl? In NaCl and in two of the four structures above, you can exchange cations and anions and still leave the structures unchanged. Which are these two structures?
3. Find out the crystal structure of palladium chloride, $PdCl_2$, and verify that it is a one-dimensional molecule.
4. What is a crystal unit cell? What are the unit cell dimensions (three lengths and three angles) for crystals of naphthalene and anthracene? Verify that all the dimensions, except one length, are practically the same for these crystals. What is the significance of the fact that $\alpha = \gamma = 90°$ in both these cases?
5. The unit cell parameters of perylene and quaterrylene are as follows: perylene, $P2_1/a$, $a = 11.27$, $b = 10.82$, $c = 10.26$ Å, $\beta = 100.55°$; quaterrylene, $P2_1/a$, $a = 11.25$, $b = 10.66$, $c = 19.31$ Å, $\beta = 100.6°$.

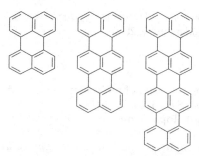

Perylene   Terrylene   Quaterrylene

   With this information, write down the space group and unit cell parameters of terrylene. This crystal structure is still unknown.
6. Refer to G. M. J. Schmidt's 1964 papers in the *Journal of the Chemical Society* given in the bibliography above. 2-Ethoxycinnamic acid crystallizes as three polymorphs α, β and γ where the Greek letters have the same meaning as they do in these papers. Write down the structures of the products that are obtained when crystals of each of the three forms are irradiated with UV light or sunlight?
7. Refer to the Cambridge Structural Database (www.ccdc.cam.ac.uk). From the CSD, retrieve the following information:

   (a) How many crystal structure determinations are there of benzene? In how many of these experiments was neutron diffraction used?
   (b) How many symmetry independent molecules are there in the asymmetric unit (Z′) for benzoic acid, naphthalene, phenol and cholesterol?

(c) How many polymorphs of the following compound are present in the CSD?

(d) How many co-crystals (multi-component molecular crystals) are there in the CSD for the anti-malarial drug artemisin?

8. Find out the crystal structure of 1,3,5,7-tetrahydroxyadamantane. Show how this structure is topologically related to the familiar CsCl structure.

9. Several workers have tried to provide up to date definitions as to what exactly a crystal is. Try and find some of these definitions. An official definition is provided by the International Union of Crystallography (www.iucr.org). From this definition, you will be able to find out why X-ray diffraction is so important in the examination of crystals.

Intermolecular Interactions

2

A molecular crystal is a periodic assembly of molecules. This regular arrangement is held together by weak intermolecular interactions that are weaker than the *intra*molecular interactions that hold atoms together — covalent bonds. So, intermolecular interactions in crystals are also called non-covalent. Crystal engineering is the theory and practice of making a molecular crystal with a particular sought after internal structure, which in turn, leads hopefully to a certain pre-desired property. Crystal engineering may therefore be broadly divided into three parts: (i) Understanding intermolecular interactions in the context of crystal packing; (ii) Developing a strategic plan by which these interactions can lead to a certain desired packing; (iii) Fine-tuning of crystal properties to achieve a pre-determined goal. The nuts-and-bolts of the entire exercise lies in a full knowledge of intermolecular interactions. This chapter describes the intermolecular interactions that are the most important from crystal engineering viewpoint.

At the outset, we must distinguish between interactions and forces between molecules. Let us first examine the figure, which shows schematically the distance profile of potential energy of a typical intermolecular interaction or, in other words, a non-bonded contact between two atoms. The energy is lowest at the equilibrium distance d_0. It is negative for all distances $d > d_0$ and also for some distances that are shorter than d_0. It is positive only for very short distances. The zero-energy line separates what we may call stabilizing (E < 0) and destabilizing (E > 0) regions. Any deviation from the equilibrium distance costs an enthalpic penalty, but this penalty is large only for large deviations in d. Interactions are described in terms of energies.

We now move from energies to forces. At the equilibrium distance, the force is zero, and the attractive

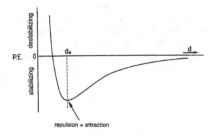

Potential energy diagram for an intermolecular interaction.

This force constant is like the force constant for a covalent bond, except that it refers to an intermolecular bond. Therefore it is much smaller.

A range of intermolecular interactions are available to a crystal engineer. Unlike crystals of extended solids that are constructed with very strong but kinetically not so accessible covalent bonds, molecular crystals are held with rather weak non-covalent interactions, which are kinetically accessible. A typical molecular crystal can be formed readily under ambient conditions. While they may be easy to prepare, they lack the stability needed for many applications. The crystal engineer often focuses on a middle ground to employ his tools. He tries to operate in a domain where both favorable kinetics and crystal stability are operative. Two approaches lend themselves to such a reality—hydrogen bonds and metal-ligand coordination bonds. The former has the advantages of versatility, reliability, and reversibility; Nature has exploited it to good advantage. Metal coordination polymers offer an additional strength with the variety of geometries they display, and the intrinsic properties of the metal ions involved.

Crystal engineering is a form of supramolecular synthesis. J.-M. Lenh, has stated that "beyond molecular chemistry based on the covalent bond there lies the field of supramolecular chemistry, whose goal it is to gain control of the intermolecular bond. It is concerned with the next step in increasing complexity beyond the molecule towards the supermolecule and organized polymolecular systems, held together by non-covalent interactions." In the present context, we should consider "supermolecule" and "crystal" as synonymous.

and repulsive components of this force are equal. For distances $d \neq d_0$, a force arises that tries to establish optimal geometry for the system. For all distances $d > d_0$, this force is attractive, and for all distances $d < d_0$, it is repulsive. The strongest attractive force occurs at the inflection point of the curve, which therefore represents quite an unstable geometry. The repulsive force becomes very large as d becomes short. The curvature at the minimum is the force constant; the sharper the minimum, the larger the force constant and the larger the forces that arise from distortions.

Throughout this book, we will use the terms "attractive" and "repulsive" to describe forces and the terms "stabilizing" and "destabilizing" to describe interactions.

2.1   General Properties

The strength of an interaction is its most obvious and notable property. Intramolecular and intermolecular interactions are distinguished by chemists, because chemistry gives great importance to the concept of a molecule, as we saw in Chapter 1. As stated at the beginning of the present chapter, it is generally presumed that intermolecular interactions are weak, in contrast to covalent bonds (*intra*molecular interactions) which are strong. However, there is no clear energetic demarcation between weak and strong interactions. Indeed, the words "weak" and "strong" are quite subjective. The strongest hydrogen bond, say the quasi-covalent symmetrical interaction in $[HF_2]^-$ anion, is worth around 50 kcal mol^{-1} while the weakest covalent bond, say the C–I bond, is worth only around 30 kcal mol^{-1}. But the majority of covalent bonds lie between 75 and 125 kcal mol^{-1}, while the majority of intermolecular interactions lie in the range 1 to 15 kcal mol^{-1}. So the chemist's distinction between "strong" covalent bonds and "weak" intermolecular interactions is generally valid. An interaction needs to be above kT in energy if it can be detected experimentally. At room temperature, this value is around 0.6 kcal mol^{-1}. Many interactions that could be important in the context of crystal packing lie in the lowest range of energies between 0.5 and 5 kcal mol^{-1}. These energies are indeed feeble. The energetic factors that go into the establishment of a crystal packing are therefore subtle, and we say that a crystal structure is the result of a compromise between the demands of many weak interactions. This is discussed in Section 2.7 for some simple examples of crystal structures that have different types of intermolecular interactions.

The second important property of an intermolecular interaction is its directionality. This characteristic is of obvious significance in crystal design because interaction directionality can be exploited to achieve specific and pre-desired intermolecular orientations. Interactions are termed as being isotropic (lacking directionality) or anisotropic (having directionality). Isotropic interactions are the ones responsible for close packing and are mainly of the dispersion–repulsion type. They include the very common C···C, C···H and H···H interactions (Section 2.7) and purely ionic interactions. Anisotropic interactions have certain extra chemical attributes that arise from specific electronic distributions around atoms. Hydrogen bonding is the most important directional interaction in molecular crystals. *It is the anisotropic character of interactions in a crystal structure that allows one to suggest design strategies for crystals of related molecules.* Let us consider, for example, the N–H···O hydrogen bond that is formed in amides. This very important hydrogen bond is found in the crystal structures of molecules of biological interest like amino acids, peptides and proteins. The hydrogen bond is generally directed in the region of the lone pairs of the acceptor O-atom. Hydrogen bonds also tend to be linear. If they are represented as X–H···Y–Z, then the angle θ around the H-atom (hydrogen bond angle) prefers to be 180°. Other hetero-atom interactions like S···S, halogen···halogen and halogen···O also have directional character. A study of many crystal structures is necessary before one can make general statements about interaction directionality. After such surveys of entire groups of crystal structures it is possible to say, for example, that the directionality of an O-atom lone pair as a hydrogen bond acceptor is better for a carbonyl O-atom, in say acids and amides, than it is for an ether or alcohol O-atom. In other words, the spread of the angle φ (angle between H, Y and Z in the hydrogen bond X–H···Y–Z) is tighter around carbonyl oxygen than it is around ether oxygen.

The third important general property of an intermolecular interaction is its distance dependence. Many non-directional stabilizing interactions have approximate inverse sixth power dependence (r^{-6}). Hydrogen bonds have electrostatic character and

How does a very weak interaction become important in determining crystal packing? Weak interactions are often numerous. A large number of them can have as considerable an effect on crystal packing as a few strong interactions. Sometimes, the effects of a large number of weak interactions may be more important than a few strong ones. This is referred to as the Gulliver effect.

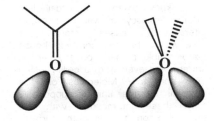

have inverse distance dependence (r^{-1}). Therefore they fall off much more gradually than the r^{-6} type dispersive interactions. Hydrogen bonds formed by weak donors and acceptors (C–H, π-bonds) have a distance dependence that lies between these extremes. The distance dependence of an interaction is relevant to the events that precede nucleation and crystallization. Electrostatic interactions like the hydrogen bond have a more gradual fall-off with distance than interactions with dispersive character. Therefore, they are viable at distances that are much longer than the van der Waals limit. Strong and even weak hydrogen bonds therefore have orienting effects on molecules prior to nucleation and crystallization. Their directional preferences are still retained in the final crystal structure even though these electrostatic interactions contribute to a lesser extent to the overall energy of the crystal.

The design of crystal structures requires a correct assessment of the energetic and spatial properties of the intermolecular interactions in it. To organise the topology or connectivity of molecules in a crystal structure, an interaction should ideally be both strong and directional. It should also be operative at long distance separations. To optimize density and packing efficiency, an interaction should be non-directional and short range. The understanding of the nature and strengths of intermolecular interactions is therefore of fundamental importance in crystal engineering.

2.2   van der Waals Interactions

A well-known example from daily life, of the efficacy of van der Waals interactions, arises from our observation that the common house lizard is able to hang on any vertical surface and even from a ceiling using only one toe. Molecules on the footpads of the animal supposedly form these interactions with the molecules on the smooth surface. Because there is a very large number of van der Waals interactions formed between these contact points, the lizard is able to suspend itself without difficulty. This is again an example of the Gulliver effect.

Let us specify what we mean by the term *van der Waals interactions*. In the context of deviations from the ideal gas laws, the term *van der Waals forces* or *van der Waals interactions* could be taken as the totality of all the attractive and repulsive forces between molecules, inclusive of hydrogen bonds. The term would always include interactions arising out of dispersion effects (induced dipole to induced dipole), the so-called London forces. It could also include interactions that arise out of attraction between two permanent dipoles or between a permanent dipole and an induced dipole. In the context of crystal engineering, the term usually means the sum of all stabilizing and destabilizing interactions, but excluding hydrogen bonding and other anisotropic interactions (such as dipole–dipole).

These isotropic stabilizing and destabilizing interactions are approximated respectively in the Lennard-Jones expression for potential energy as the inverse sixth power and inverse twelfth power terms. The term *van der Waals interaction* is used in this context, in this book.

Interactions of the van der Waals variety are, at an individual level, relatively weak but, as mentioned earlier, they are very numerous. Therefore, when taken as a whole, they account for a substantial amount of the crystal stabilization (sublimation energy). Accordingly, they account for many characteristics of organic compounds, such as solubility, density and melting point. They always increase with an increase in the non-polar content of the molecule.

As mentioned above, van der Waals interactions arise from attractive (dispersive) and repulsive forces. These interactions are isotropic in nature. Dispersion leads to an interaction between fluctuating monopoles in adjacent molecules. The magnitude of the interaction is roughly proportional to r^{-6} where r is the interatomic separation and is approximately proportional to the size of the molecule. The repulsive forces have approximate r^{-12} distance dependence. They are stabilizing at distances close to the equilibrium separation but can become rapidly destabilizing at shorter distances. Attractive and repulsive forces balance each other in a stable crystal. Most interatomic separations in a crystal are close to the equilibrium distance. The majority are a little longer than this equilibrium value. A few are shorter.

2.2.1   *Close Packing*

The trade-off between these sets of stabilizing and destabilizing interactions with some contribution from electrostatic interactions is the cornerstone of the atom-atom potential method of A. I. Kitaigorodskii to describe molecular crystals. This model describes all crystals as being derived by the most efficient utilization of space, in other words from closest packing. Accordingly, it is called the *principle of close packing*.

The close packing principle states that molecules pack in a crystal so that any reference molecule is surrounded by a maximum number of close neighbors. Generally, this number is 12 so that if a molecule is approximated by a sphere, the crystal structure will approximate to a cubic close packed (*ccp*) or hexagonal

Kitaigorodskii, conducted an ingenious experiment that showed that the repulsive part of the potential energy curve rises much more steeply from the equilibrium distance than the attractive part. He showed that the solubility of anthracene in acridine (or 9-azaanthracene) is much less than that of acridine in anthracene. The individual crystal structures are distinct and his result means that it is easier to introduce a smaller molecule into a matrix of a larger one, than the other way round. In other words, one pays a smaller energy penalty to create hollows in a crystal structure than one does to squeeze molecules into an already full structure.

Lennard-Jones expression for the potential energy of a system of two non-bonded atoms i and j

$U_{ij} = -Ar_{ij}^{-6} + Br_{ij}^{-12} + q_1q_2r^{-1}$

U_{ij} Potential energy between two atoms at a distance r_{ij}

A, B are constants

The first term is dispersive (attraction)

The second term arises from repulsive forces

q_1 and q_2 are charges on the two atoms and the third term is an empirical representation of the electrostatic interactions.

The beauty of Kitaigorodskii's model is that it explains why only a handful of the 230 theoretically possible space groups are actually observed for molecular crystals. Organic molecules are irregularly shaped. Therefore the cubic and hexagonal crystal systems, which are observed for the *ccp* and *hcp* arrangements, are of too high a symmetry for most organic crystals. Most common is the monoclinic system which is the best compromise between an irregular molecular shape and the need to have translational symmetry elements, which favor close packing. The most common space group for molecular organics is $P2_1/c$ in which a dissection along any principal direction yields close packed layers of molecules. This single space group is adopted by as many as 30% of all organic crystals. The equivalent space group for resolved compounds is the monoclinic space group $P2_1$ which is adopted in 10% of the cases. The fact that these space group preferences also extend to crystals with anisotropic interactions is interesting and is a matter for more detailed study.

close packed (*hcp*) arrangement. The model assumes that molecules in a crystal tend to assume equilibrium positions so that the potential energy of the system is minimized. Since the attractive and repulsive forces between molecules are assumed to be isotropic, molecules tend to approach each other so that the number of lowest energy interactions is as large as possible. In other words, the number of intermolecular contacts in a crystal tends to a maximum and these contact distances cluster around distances associated with energy minima. The twin requirements of a maximal number of contacts with a narrow distance spread means that molecules pack in ways so that the projections in one molecule dovetail into the hollows of its neighbors. Ideally, this dovetailing occurs in three dimensions so that a close packed structure yields close packed two-dimensional layers in several dissections. In each of these layers, a reference molecule is coordinated optimally by six others.

The close packing principle generally holds well for hydrocarbons, because the intermolecular interactions in hydrocarbon crystals (C⋯C, C⋯H, H⋯H) are all isotropic. Condensed aromatic hydrocarbons, $C_m H_n$ provide a good example of the application of this principle. The packing arrangements of these molecules are governed by tendencies towards stacking in the carbon rich molecules (say $m/n \geq 2$) and towards inclined herringbone geometries in hydrogen rich molecules (say $m/n < 1.5$). The carbon to hydrogen stoichiometry is thus a critical parameter in these crystal structures. Typical examples of the stacked and

Most frequently observed space groups for organic crystals in the CSD (April 2011). Compounds contain at most C, H, N, O.

Space Group	Symmetry Elements	Frequency of Occurrence
$P2_1/c$	screw, glide	10983
$P\bar{1}$	inversion	9868
$P2_12_12_1$	screw	8566
$C2/c$	2-fold, glide, inversion	3020
$P2_1$	screw	6084
$Pbca$	glide, inversion, screw	2052
$P1$	identity	958
$Pna2_1$	screw, glide	752
$Pbcn$	glide, inversion, screw	320

herringbone geometries are provided by coronene, $C_{24}H_{12}$, and naphthalene, $C_{10}H_8$. Long hydrocarbon chains in aliphatic compounds also pack in a close packed manner. When viewed in section, they appear as would a section of an *hcp* or *ccp* layer. This packing is called the hydrophobic effect and is seen in unbranched hydrocarbon chains longer than five carbon atoms. This effect is the basis for forming layers of monomolecular in Langmuir-Blodgett films.

Close packed structures are based on the efficient packing of irregularly shaped objects. The structures that are obtained are characterized by interlocking of bumps and hollows in neighboring molecules. Interactions between the molecules are isotropic and are of the dispersion-repulsion type. They do not depend so much on the chemical nature of the interacting atoms as long as the numbers of electrons in the contacting atoms are comparable. Accordingly, molecules of a similar shape and size take the same or similar crystal structures. Benzene and thiophene form such an isostructural pair of molecules. Many chloro and bromo derivatives of the same molecule (say 1,4-dichlorobenzene and 1,4-dibromobenzene) adopt the same crystal structure. Even more interesting is the fact that analogous chloro and methyl derivatives of a hydrocarbon adopt the same crystal structure. The chloro and methyl groups are equivalent from geometrical considerations (both are roughly spherical and their volumes are 19 and 23 Å^3 respectively) but they are chemically distinctive. Therefore, if chloro and methyl derivatives of a compound are isostructural, it means that geometrical factors are important in the crystal packing, and not chemical ones. Very often these isostructural compounds form solid solutions with one another. This indicates that the interactions between molecules in the pair are indeed very similar.

It is a universal tendency in solids that the requirements of close packing and interaction directionality work at cross purposes to one another. Close packing means that there is a driving force for any reference molecule to be surrounded by a maximum number of near neighbors. Any particular interaction between a particular pair of atoms is not so important. Interaction directionality means that the orientational demands of a particular anisotropic interaction need to be taken into account in arriving at the crystal structure. Because of this, close packing could be

Two-dimensional closest packing as seen in M. C. Escher's "Horsemen". For molecules to come together in this way, all portions of their surfaces must be equally "sticky". The presence of a few "stickier" regions leads to distortions from ideal close packing, and eventually to a breakdown of the geometrical model.

A. I. Kitaigorodskii with some of his associates in his laboratory in Moscow (ca. 1960).

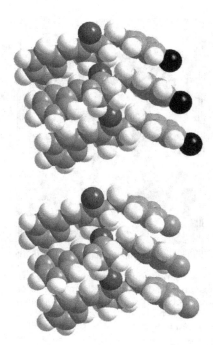

Isomorphous chloro and methyl derivatives of 5-benzyl-2-benzylidene-cyclopentanone. In these compounds, the molecular framework is large and non-polar. The roles of the chloro and methyl groups on the crystal structure are the same, namely in space filling and close packing. The crystallographic unit cells are practically identical. Such *chloro-methyl exchange* is strong evidence of crystal structures that are governed by close packing. A similar tendency is seen in pairs of compounds in which a CH=CH fragment is replaced by a thioether linkage, –S–, the so-called *benzene-thiophene exchange*.

By geometrical factors are meant those arguments and reasons that are derived from molecular shape and size. But, in the end, all these factors have a fundamental chemical origin. The shape of a molecule in a crystal is related to the repulsive potential field around it.

compromised. In the inorganic world, close packed or directionless structures are exemplified by the crystal structures of the metallic elements and noble gases. Directional structures are exemplified by diamond, ZnS and ice. In close-packed structures, the coordination number around a reference molecule (number of closest neighbors) strives to a maximum of 12. In directional structures, the coordination number can be as low as 4. These tendencies are also seen in organic molecular solids but not in such an obvious way, because the interactions are quite weak. Structures generally tend towards close packing but there are local deviations because of directional interactions.

When hetero-atoms are present in the molecule, there is the possibility of the formation of anisotropic interactions. These interactions cause deviations from pure close packing as envisaged by Kitaigorodskii, to a greater or lesser extent. The most important anisotropic interaction in organic crystals is hydrogen bonding.

2.3 Hydrogen Bonds

Hydrogen bonding has been called the master-key of molecular recognition. It is the most reliable interaction in the toolkit of the crystal engineer. It is both strong and directional. In a milieu where most of the interactions are weak and non-directional, in other words of the van der Waals type, strong and well directed interactions can have important effects on the establishment of a stable crystal structure.

The typical hydrogen bond can be represented as an interaction between a donor X–H and an acceptor Y–Z. The hydrogen bond is then written as X–H···Y–Z with the three dots signifying the bond. Scientists have been studying hydrogen bonds for just under 100 years. In the early days of the subject, it was assumed that the elements X and Y are highly electronegative non-metals like F, O or N. Only in such a case was it felt that the X–H could become sufficiently polarized so that it would be attracted electrostatically to the electronegative Y atom. So a hydrogen bond was shown as $X^{(\delta-)}–H^{(\delta+)}\cdots Y^{(\delta-)}–Z$. It was always known, however, that there was some small amount of covalent character in a hydrogen bond. This is because it was observed that in the infrared spectrum, the stretching frequency of the X–H bond was depressed to lower energies upon hydrogen bond formation. This bathochromic shift (shift to a lower frequency) in other words a weakening of the X–H covalent bond was taken as an indication of covalent character in a hydrogen bond. In some cases, like the hydrogen bifluoride ion, $[HF_2]^-$, the hydrogen atom lies equidistant between two electronegative fluorine atoms and the hydrogen bond is quasi-covalent. In fact, the two H–F bonds in this anion are equivalent and each has a quasi-covalent character.

Along with spectroscopy, X-ray crystallography was another experimental technique that brought about great improvements in our understanding of the hydrogen bonding phenomenon. With crystallography, the chemist could obtain accurate estimates of the length and angle properties of a hydrogen bond. Generally, it was observed that in a hydrogen bond of the type X–H···Y–Z, the distance between the elements X and Y was much shorter than the sum of their van der Waals radii. The X to Y distance is referred to as D, while the H to Y distance is referred to as d. Take for example, a typical and well known hydrogen bond like N–H···O=C that is found in amides, amino acids, peptides and proteins. The typical N···O non-bonded distances in such hydrogen bonds, namely D values, lie in the range of 2.7–2.8 Å. The van der Waals radius of N is 1.5 Å, while that of O is 1.6 Å. So there is a shortening of around 0.3 Å in the formation of the hydrogen bond. From the early days of the subject, such a shortening was taken as a sure indicator of hydrogen bonding. Later on, it was possible to determine the positions of H-atoms in a crystal structure with crystallography, and the shortening of the H···Y distance d was measured with respect to the van der Waals sum for the elements H and Y. Here too, a shortening was seen in many cases where the spectroscopic evidence indicated the formation of a hydrogen bond. Depending on the nature of the elements X and Y, typical distance ranges were obtained for various hydrogen bonds.

The angular properties of hydrogen bonds are also important features of the interaction. The tendency for all hydrogen bonds is towards linearity, in other words, the X–H···Y angle, also referred to as the hydrogen bond angle or θ, tends towards 180°. This is because the effectiveness of the positively charged H-atom as an agent that can screen the negative charges on the closely approaching electronegative

The packing coefficient, C_{pack} in a crystal structure can be obtained using the relation,

$$C_{pack} = Z.V_{mol}/V_{cell}$$

where V_{mol} = molar volume, V_{cell} = unit cell volume and Z = number of molecules in the unit cell. The average value of C_{pack} for organic crystals is around 0.7.

An early definition (1939) of the hydrogen bond X–H···Y is by Linus Pauling who stated that "under certain conditions an atom of hydrogen is attracted by rather strong forces to two atoms instead of only one, so that it may be considered to be acting as a bond between them." Instead of being attracted only by X, the hydrogen atom is attracted to both X and Y.

A current definition (2010) of the hydrogen bond states that it is "an attractive interaction between a hydrogen atom from a molecule or a molecular fragment X–H in which X is more electronegative than H, and an atom or a group of atoms in the same or a different molecule, in which there is evidence of bond formation." Notice that unlike Pauling's definition, the modern definition makes no assumptions about strong forces but rather refers to "evidence of bond formation" without specifying what evidence is satisfactory and without stating what a bond is. This relatively open definition has been recently accepted by the International Union of Pure and Applied Chemistry. The three most important features of this new definition are: (i) The hydrogen bond is attractive; (ii) H is more electropositive than X, the atom to which it is covalently bonded; (iii) Evidence of bond formation is needed.

Hydrogen bonds, X–H···Y–Z				
Strength	**Examples**	**X–Y** (**D**, Å)	**H···Y** (**d**, Å)	**X–H···Y** (**θ**,°)
very strong X–H ~ H···Y	[F–H–F]⁻	2.2–2.5	1.2–1.5	175–180
strong X–H < H···Y	O–H···O–H O–H···N–H N–H···O=C N–H···O–H N–H···N–H	2.6–3.0 2.6–3.0 2.8–3.0 2.7–3.1 2.8–3.1	1.6–2.2 1.7–2.3 1.8–2.3 1.9–2.3 2.0–2.5	145–180 140–180 150–180 150–180 135–180
weak X–H << H···Y	C–H···O	3.0–4.0	2.0–3.0	110–180

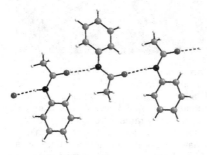

The tendency of hydrogen bonds to be linear is satisfied admirably in the crystal structures of secondary amides. Here, the N–H···O=C bonds are linear and connect the molecules in a one-dimensional array. The illustration here shows the crystal packing in acetanilide.

An interesting way in which hydrogen bonding and van der Waals interactions may be used to design well formed crystals, is to co-crystallize a molecule that can donate hydrogen bonds with triphenylphosphine oxide, an excellent acceptor of hydrogen bonds. A phenol, say can form a hydrogen bond of the type O–H···O=P in the binary crystal that is formed. The presence of three phenyl groups on the phosphorus atom allows the binary compound to form good crystals. Growth rates in all directions are comparable and good because of the herringbone C···H interactions that can form in three dimensions. This technique is useful for compounds that do not easily give good crystals on their own.

atoms X and Y is at a maximum when the H-atom lies on the straight line that connects the atoms X and Y, in other words when $\theta = 180°$. This linear geometry is not always possible in crystal structures although it may be easier to achieve this in gas phase aggregates. Any crystal structure is a result of compromise between many intermolecular interactions of different strengths, angular preferences and distance dependence characteristics. Therefore a hydrogen bond (or any other interaction for that matter) is affected by all the other interactions in the crystal structure. Strong hydrogen bonds like N–H···O and O–H···O are the least affected by other interactions in the crystal, because they are strong. They are more nearly linear and their distance distributions cluster around small ranges. Weak hydrogen bonds, say C–H···O or O–H···π, are more likely to be bent. They also occur over wider distance ranges. Very weak hydrogen bonds like C–H···π can have very non-specific geometries both with regard to length and to angle properties.

Our original ideas about the hydrogen bond date from the time of Pauling, who envisaged the interaction as being formed by two rather electronegative atoms X and Y. The strength and accordingly distinctive directional character of the interaction were attributed as arising from the fact that the electronegativities of X and Y caused a considerable amount of electrostatic character to the H···Y interaction. Gradually, questions began to be raised about the nature of the interaction X–H···Y–Z when the electronegativities of either or both X and Y were moderate to low. Did the interaction retain the "characteristic" properties of a hydrogen bond in interactions like C–H···O, C–H···N,

O–H⋯π, N–H⋯π, C–H⋯π, O–H⋯Tr (Tr = transition metal), M–H⋯O (M = any metal)? Answers to such questions were initially slow in coming but a great deal of experimental and theoretical evidence in the last two decades has shown that these weaker interactions retain many significant properties of the strong hydrogen bonds; most significantly, and in the crystal engineering context, their effects on crystal structure and as determinants of stable crystal packing, weak hydrogen bonds can be nearly as effective as their stronger counterparts.

The hydrogen bond is a predominantly linear interaction ($\theta = 180°$), and yet the configuration of hydrogen bonds usually observed in crystal structures is rarely linear because the number of possible spatial configurations with the X–H⋯Y angle in the range θ to $(\theta + d\theta)$ is proportional to $\sin \theta$. In practice, the median value for θ is around 165°. A further reason for such bent approaches is that the hydrogen atoms are often approached by a second acceptor in an attractive fashion. This type of interaction is called a *bifurcated hydrogen bond*, or a three-centered interaction. This terminology is used to contrast bifurcated hydrogen bonds from simple hydrogen bonds X–H⋯Y–Z that have a single donor and a single acceptor; such hydrogen bonds are called two-centered. The two centres are the atoms X and Y. Occasionally, three acceptors may form non-covalent interactions with a hydrogen atom and this would be called a trifurcated or four-centered bond. The four centres are the four electronegative atoms that surround the hydrogen atom, one covalently and three non-covalently.

Multifurcated arrangements are common in organic crystals because the typical organic molecule has many acceptor groups but only a few donor groups (OH and NH_2). There is usually an excess of acceptors over donors. So rather than form a few simple (two-centered) hydrogen bonds and leave many acceptors completely unsatisfied, the molecules prefer to adopt multifurcated geometries — one donor and many acceptors. The converse is the multifurcated acceptor — a single acceptor atom is surrounded by several donors. How does this happen in a donor-poor situation? Generally, one of the donors is of the "strong" type, O–H or N–H while the other is of the "weak" type, usually C–H, of which there are many in the molecule. In still other cases, acceptor-rich molecules crystallize as hydrates. Water is perhaps the most

A hydrogen bond, represented as X–H⋯Y–Z, is an electrostatic interaction between an electropositive hydrogen atom and two electronegative atoms X and Y. However, there are also other features of this interaction: notably there is a small amount of covalency and also some dispersive-repulsive character. Covalency is more important when very powerful donors and acceptors are involved. The van der Waals nature of the interaction is more pronounced when the donors and acceptors are weak, that is when the electronegativities of X and Y are subdued. In extreme cases of repulsion, hydrogen bonding can even lead to an increase in the IR stretching frequency of the X–H bond, the so-called blue shifted hydrogen bonds. This occurs for extremely weak acceptors such as C–F. But all these varieties of interactions are still called hydrogen bonds.

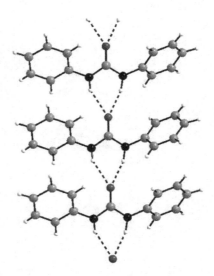

Ureas are simple organic molecules in which strong donors outnumber the acceptors. The symmetrical diphenylurea for example (two donors, one acceptor) forms a tape structure in which the acceptor oxygen atom is bifurcated. It receives hydrogen bonds from two equivalent donor groups.

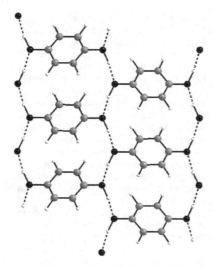

The crystal structures of phenols display a rich diversity of cooperative patterns. The packing of γ-hydroquinone is a typical example of a cooperative arrangement of hydrogen bonds arranged around a 2_1 screw axis.

important molecule in which the donors outnumber the acceptors. So a hydrated system has a better chance of forming more two-centered interactions than does an anhydrous compound.

Hydrogen bonding is usually associated with the property of cooperativity, and this property is pronounced in organic crystal structures. Polarizability and charge transfer effects make the total binding energy of all hydrogen bonds in an aggregate greater than the energy sum of the individual hydrogen bonds. Take for example an O–H group in an alcohol, ROH. When ROH forms a hydrogen bond with another such molecule, the acidity of the free hydrogen atom in the O–H···O–H dimer is greater than the acidity of the hydrogen atom in an isolated O–H group. So the dimer forms a hydrogen bond easier than a monomer. Likewise a trimer forms a hydrogen bond easier than a dimer. So there is a tendency for aggregation that increases with the size of the aggregate. This tendency does not continue indefinitely, however. In practice, the cooperative effect tails off after six or seven hydrogen bonds are formed. In summary, hydrogen bonds are more stable when they are involved in dimers, trimers, chains and infinite two- and three-dimensional structures. All this is of importance in crystal growth as well. Patterns of cooperative interactions are particularly well favored on kinetic grounds and are found often in organic crystals.

2.3.1   *Weak Hydrogen Bonds*

In weak bonds X–H···Y–Z, the electronegativities of the donor and acceptor atoms X and Y are reduced. Typically X is carbon. Y may be a multiple bond like C=C or C≡C or an aromatic ring. In contrast to strong hydrogen bonds, weak hydrogen bonds are compressed or expanded and bent or straightened by other crystal packing forces. The weaker the hydrogen bond, the less discernible are the associated directional characteristics. C–H···O bonds are commonly encountered in organic crystal structures owing to the frequent occurrence of the oxygen atom in simple organic molecules. These interactions can play a significant role in crystal packing.

Like other hydrogen bonds, C–H···O interactions have electrostatic character with a long range distance fall-off. The length of a C–H···O bond depends on both the acidity of the C–H group and the basicity of the oxygen atom. The more acidic C–H groups form shorter C–H···O hydrogen bonds. The more weakly acidic C–H donors form long hydrogen bonds, whose length may exceed the van der Waals separation distance. In practice, many longer C–H···O contacts (C···O 3.50–4.00 Å) have angular characteristics and effects on crystal structures that resemble the shorter contacts (3.00–3.50 Å). The C–H···O bond is not a van der Waals contact but is primarily electrostatic, falling off much more slowly with distance and hence viable at distances that are equal to or longer than the van der Waals limit. Even long C···O separations (~4.00 Å) may need to be considered; these contacts may have their origin in certain preferred orientations of molecules as they approach each other during crystallization. Accordingly, the study of weak hydrogen bonds is important in crystal engineering.

Typical C–H···O hydrogen bond angles (θ) occur in the range 100–180°. They cluster around 150–160° for reasons stated earlier. The length of a C–H···O bond correlates inversely with the angle. This is an essential criterion for hydrogen bonding. Most strong hydrogen bonds cluster in small length and angle ranges. Weak hydrogen bonds like C–H···O are found in larger length and angular ranges because they are weak: they can be distorted by other forces in the crystal.

While C–H···O hydrogen bonds are widespread, the complementary O–H···π interactions are rare because carbon is not as electronegative as oxygen and also because carbon atoms are not often situated in sterically unhindered positions, unlike carbonyl and ethereal oxygen atoms that permit easy access by C–H groups to form C–H···O hydrogen bonds. C–H···π interactions are even weaker than O–H···π and C–H···O hydrogen bonds because both the donor and acceptor fragments are weakly polarized. There is no well accepted convention in which C–H···π interactions are accepted as hydrogen bonds. When the donors and acceptors are somewhat stronger, as in say acetylene and other alkynes, the definition of a C–H···π geometry as a hydrogen bond is somewhat less contentious.

2.3.2 *Hierarchies of Hydrogen Bonds*

The hydrogen bond is both strong and directional. In almost every crystal structure of a molecule that contains hydrogen bonding functionalities, the packing cannot be understood without taking into account the hydrogen bond patterns. Hydrogen bond donors like

Benzoquinone is a good example of a molecule that forms C–H···O hydrogen bonds in the crystal structure. The C–H groups are activated because they are conjugated with a C=O group, and the carbonyl oxygen atoms are good acceptors of hydrogen bonds. The melting point of benzoquinone (108°) is surprisingly high and, in part, this could be because of C–H···O hydrogen bonding in the solid.

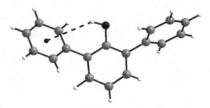

In 2,6-diphenylphenol, the acidic hydrogen atom cannot be approached by the oxygen atom from a neighboring molecule because it is blocked by the *ortho*-phenyl groups. O–H···O–H hydrogen bonding is not possible. The molecule forms an intramolecular O–H···π hydrogen bond.

OH, NH, and NH$_2$ are always used unless the hydrogen atom is in a sterically hindered location. Hydrogen bond acceptors like OH, C=O, NH$_2$, –N= and –O– are often used. We have explained in the previous section when and how the crystal structure accommodates the usual surfeit of acceptors over donors in a typical organic molecule.

In compounds that contain multiple hydrogen bond donor and acceptor functional groups, these groups can pair up in the hydrogen bonds that are formed according to their strength. So, the best hydrogen bond donor forms a hydrogen bond with the best hydrogen bond acceptor, the second best donor with the second best acceptor and so on. Clearly, there will be many exceptions to this generalization but it is still a useful guide. Consider, for example, isonicotinamide. In this well studied molecule, there are two basic sites that can act as hydrogen bond acceptors. These are the pyridine nitrogen atom, which is the more basic acceptor, and the amide carbonyl oxygen atom, which is the less basic site. When two different aromatic acids are combined with isonicotinamide in a 1:1:1 ratio, the pyridine site forms a hydrogen bond with the stronger carboxylic acid, while the amide functionality, the second best acceptor binds to the weaker acid.

Among the reasons why such hydrogen bond hierarchies are not followed in crystals are: (i) The numbers of donors and acceptors are quite different, leading to multifurcation and hydration; (ii) The donor and acceptor groups are located in close proximity to one another in a molecule, and donor and acceptor strengths need not be a reliable guide to predicting the hydrogen bonds that could be formed. Steric factors may also need to be taken into account; (iii) The compound is polymorphic and different hydrogen bonds can be formed in the different morphs. A hydrogen bond hierarchy that is based on donor and acceptor strengths is a characteristic of kinetic crystallization. Under conditions when thermodynamic crystallization takes place, these hierarchies may not apply. This is discussed in more detail in Section 5.4.

2.4   Halogen Bonds

It has long been known that the halogens form short non-covalent contacts in crystals. In the past, the nature of these short contacts was generally obscure and unknown. More recently, it has been found that many of these short halogen atom contacts arise from polarization. A halogen atom, which is covalently bonded to a carbon atom, is polarized positively in that region of the atom that is furthest away from the carbon atom. This is called the polar region. The equatorial region is then polarized negatively because the overall charge on the atom is zero. In this way, a halogen atom is able to make an electrostatic contact with a negatively polarized or electronegative species, if it approaches the latter via its polar region. This contact is reminiscent of a hydrogen bond in the sense that a positively charged species (halogen instead of hydrogen) approaches a negatively charged species. Accordingly, such an interaction has been termed as a *halogen bond*.

The polar region of a halogen atom covalently bonded to carbon is polarized positively while the equatorial region is polarized negatively.

The term *halogen bond* is of recent origin. Interactions of the type I···N, I···O, Br···O are, however, very well known. In the 1960s, Odd Hassel (who won a Nobel Prize for this work) showed that the 1:1 molecular complex of bromine and dioxane was assembled with polarization induced $Br^{\delta+}···O^{\delta-}$ interactions. This would be called a halogen bond today, because it involves electropositive halogen. Notice that contacts involving the heavier halogens (Br, I) were studied first. Polarization effects are expected to be more pronounced in these cases; there is no doubt that halogen bonds have a considerable component from polarization. In later years, halogen bonds that involve Cl were also studied.

Halogen bonds can be unsymmetrical (X···Y, where Y can be a halogen or non-halogen atom) or symmetrical (X···X). Unsymmetrical halogen bonds involve the approach of an electropositive halogen atom to an electronegative atom (halogen or otherwise). Contacts like I···Br, I···Cl, I···N, I···O, Br···O and Cl···N belong to this category. Symmetrical Cl···Cl, Br···Br and I···I contacts can be sub-divided into two geometrical categories. In the first (called Type-I), the C–X bonds in the interaction C–X···X–C are exactly or nearly parallel. These contacts cannot strictly be called halogen bonds because crystal symmetry, or pseudosymmetry, dictates that electrostatically identical regions of the adjacent halogen atoms make the closest approach. The second category (Type-II) is a genuine halogen bond. The two C–X bonds make an angle of nearly 90° with respect to one another. In this geometry, the electropositive region of one halogen atom can make a close approach to the electronegative region of the neighbor.

The study of halogen bonding is under active consideration today and much more can be expected in terms of its use as a design element in crystal engineering.

2.5   Other Interactions

Like the halogens, polarization effects are important for sulfur, which is known to form short directional contacts of the type S···N, S···S and S···Cl. In all these cases, the sulfur becomes electropositive. The S···S contact is polarized as $S^{\delta+}\cdots S^{\delta-}$. Intermolecular contacts formed by the other chalcogens Se and Te, as well as other metalloids like Bi have been identified. In many of these cases, charge separation leads to attraction between non-bonded atoms often with considerable directionality and orientational preferences.

Metal atom interactions have also been examined as design elements in crystal engineering. Metals, of course, are integral elements of coordination polymers but in those cases, the primary interaction is a coordinate bond between a metal cation and a negatively charged ligand. This does not qualify as an intermolecular interaction in the strict sense of the term. More pertinent in the context of non-covalent interactions that involve metals are the Au···Au and Ag···Ag interactions. The aurophilic Au···Au interaction is well documented and its use in crystal engineering has been recorded. Whether the corresponding argentophilic Ag···Ag interaction may be reliably used in design strategies is still an open question. Metal atoms can act as hydrogen bond donors and as hydrogen bond acceptors. Halogen atoms attached to metal atoms (M–X) can act as powerful hydrogen bond acceptors.

Interactions between ions have been studied from the crystal engineering viewpoint. However, there is not much systematic work in this area. It is important to distinguish between the larger organic cations and anions and the smaller inorganic ions. The effects of these two types of ions on the crystal packing of a molecular solid are different. Large polarizable organic cations and anions are rather similar to neutral molecules in terms of their effects on crystal packing. Smaller inorganic ions can have considerable effects in directing crystal packing. Cations like ammonium, guanidinium, amidinium and anions like carboxylate and sulfonate are often utilized because the crystal engineer can take advantage of the strong charge-assisted interactions that ensue. Metal cations are constituents of coordination polymers and their large and important subset, the metal organic framework (MOF) compounds. These are discussed in Chapter 7.

2.6   Methods of Study of Interactions

We will now review the various methods of probing the presence and nature of intermolecular interactions in molecular solids. Typically, there are three broad ways of doing this. One can use crystallography to image the internal structure of a crystal. Crystallography yields information on the geometrical properties of interactions such as lengths and angles. When large amounts of crystallographic data are mined, typically from databases, additional information, including information of a chemical type, can be extracted. The second approach which one may use to examine interactions is through spectroscopy. Spectroscopy gives information on the energies of interactions. Information from these two experimental techniques, namely crystallography and spectroscopy, feeds into the third broad strategy, that is computation. Computation provides information on both geometrical and energetic properties of intermolecular interactions. The results of computation ideally feed back into the design of new and better experiments both of a crystallographic and spectroscopic variety.

2.6.1   *Crystallography*

Crystal engineering deals only with crystalline solids. Crystallography is an important experimental technique in this subject because it deals with diffraction, and crystals diffract beams of X-rays, neutrons and electrons. The internal structure of a crystal can be imaged with X-ray diffraction. When

Anti-crystal engineering

Crystal engineering deals with diffracting crystals, and strictly speaking will exclude the study of amorphous solids. However, there are areas of the subject that are concerned with amorphous solids. The most important among these is the study of drugs and active pharmaceutical ingredients (API). Drugs and APIs often exist in many polymorphic modifications and some of these may also be amorphous. The latter compounds may often be more soluble than their crystalline counterparts and, as such, more desirable from a medicinal viewpoint. But amorphous forms are also generally less stable than crystalline polymorphs. Therefore, it might be advantageous to design molecules that will provide stable amorphous forms. This has been referred to as "anti-crystal engineering."

Solving a crystal structure means obtaining a model of atom connectivity that makes chemical sense from the diffraction intensities. In technical terms it consists of assigning correct phases to experimental diffraction intensities that are phaseless. In the context of small molecule crystal structures, which are normally the concern of crystal engineers, the approach of choice for phase assignment is referred to as Direct Methods. This approach uses reflections that are intense when weighted for the interplanar spacings to which they correspond. Typically, no more than a few hundred reflections are used for solving the structure, out of a data set that may consist of several thousand diffraction intensities. Solving a structure is different from *refining a crystal structure.* In refinement, the model that is obtained after phase assignment is compared to the entire set of diffraction data and the atomic parameters are allowed to adjust themselves slightly so as to minimize the difference between the observed diffraction data and values calculated from the structural model. Both crystal structure solution and refinement are nearly automatic procedures with present day computers.

the domains of the ordered, periodic regions of a crystal are sufficiently large, we have a *single crystal.* Single crystals may be examined with monochromatic X-rays and the diffraction patterns that are obtained are analyzed to get a direct image of the atomic positions in the crystal. This task is a practically routine operation today for most crystals of organic molecules that contain less than say 100 atoms per molecule. The instrument that is used to collect the diffraction data is known as a single crystal diffractometer. When the crystalline domains are smaller than a critical size, one has a *microcrystalline solid*, also referred to as a *powder.* Such a solid, which has crystallites in many orientations, will also give a diffraction pattern and this is recorded with an instrument known as a powder diffractometer. The information that is obtained in a powder diffractogram is significantly less than what is obtained from a single crystal diffractometer. Getting the atomic positions in the crystal structure from powder data is a non-routine task at the present time. While single crystal diffractometry yields accurate structural information practically automatically, this is not true of powder data which may be generally used only as a fingerprinting device to characterize a crystalline material. Finally, when the crystal sizes become exceedingly small and when these sizes occur over ever broadening ranges, we reach a stage when sharp diffraction peaks are no longer observed. These solids are termed as *X-ray amorphous.* In contrast, we have *glasses* where any kind of long range periodicity is absent. Amorphous solids do not diffract X-rays and so the methods of crystallography cannot be applied to them.

Diffraction methods are especially important to the study of intermolecular interactions. There are three main reasons for the rapid growth and development of X-ray and neutron crystallography: (i) A near automation has been achieved in solving crystal structures and in refining the model so obtained with respect to the experimental diffraction data; (ii) Powerful and yet inexpensive computing facilities are available practically everywhere; (iii) Constant improvement and also miniaturization of diffractometers, including detector technology, has led to a significant improvement in time required to obtain diffraction data from a single crystal. These three reasons are now elaborated further.

Diffraction data for crystal structures with less than 100 non-H atoms in the asymmetric unit may now be phased routinely. Refinement of the model is practically immediate. PCs and clusters are routinely available and the results of crystal structure determinations may be analysed and displayed in a number of aesthetically pleasing ways. Access to the internet means that a large amount of public domain software may be gainfully employed for the understanding and dissemination of crystallographic knowledge. Programs that are used to solve and refine crystal structures produce outputs in standard formats (such as the CIF or Crystallographic Information File) that will interface with all kinds of other software. Even the publication process has come under the sway of these computational advances and *Acta Crystallographica* was the first journal to demand submissions exclusively in an electronic format for refereed contributions. As for diffractometers, the first cycle of fully computer-controlled four-circle instruments, collecting diffraction data one at a time using point detectors, began around 1975. This era lasted for 25 years, till around 2000, and has now completely given way to charge coupled device (CCD) diffractometers. These instruments are based on image-plate detectors and can record diffraction intensities of many reflections simultaneously, and the time required for a data collection has shrunk by an order of magnitude. The latest advances have to do with table top diffractometers that can collect crystal data in 1–3 hours as opposed to the first CCD machines that required around 6–12 hours. All this means that a very large number of accurate crystal structures are being determined today. In 2000, there were around 200,000 published crystal structures of organic and organometallic compounds. Today there are more than 500,000.

Neutron diffraction is particularly useful for the study of hydrogen bonds because with this technique the crystallographer is able to establish very accurately the position of the hydrogen atom, which diffracts poorly with X-rays. This technique is of importance in the study of hydrogen bonding to multi-atom π-bases, where it is not possible to infer the hydrogen atom positions on the basis of geometrical considerations.

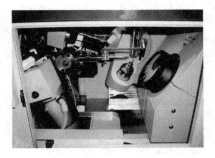

A modern table top single crystal diffractometer. The image plate detector is the cylindrical disk on the right (courtesy, Rigaku Corporation).

2.6.2 *Crystallographic Databases*

An individual crystal structure determination gives geometrical information on that particular structure.

The Cambridge Structural Database is very useful in crystal engineering. Apart from its utility in studying the properties of intermolecular interactions, here are some other applications:

(i) To find out if a particular crystal structure has already been determined.

(ii) To follow trends in crystal packing, say an examination of the number of molecules in the crystal asymmetric unit, Z'.

(iii) To examine more closely the phenomena of polymorphism and pseudopolymorphism.

(iv) As a knowledge bank for crystal structure prediction.

When the results of structure determination of many crystals are archived together, one gets a crystallographic database. There are many situations in which the results of structure determinations of a collection of many crystals provide information that would not be obtained from a single structure determination. Databases are particularly useful in crystal engineering because the packing features of whole groups of structures can be examined. Databases are also useful to probe the properties of intermolecular interactions.

The Cambridge Structural Database (CSD) which is produced by the Cambridge Crystallographic Data Centre, CCDC, (www.ccdc.cam.ac.uk) is a depository of X-ray and neutron data on organic and organometallic crystals. Starting with less than 2000 entries in 1965, the CSD today contains accurate X-ray crystal structures of over a half million compounds. The CSD has built-in, user-friendly data retrieval, statistical analysis and crystal structure display and visualization packages. The output from a particular query may be compared with a knowledge base of intermolecular interactions. With the CSD, chemists and crystallographers have convenient access to a large amount of crystallographic data and structural information. The CSD is the crystallographic database that is the most relevant to the subject of crystal engineering. Databases can be used to perform a number of tasks in structural chemistry. In this section, we will describe the use of the CSD in studying intermolecular interactions. This is best done by looking at a few examples.

In the accompanying figure, we see d-θ distance-angle scatter plots for strong (O–H···O) and weak (C–H···O) hydrogen bonds in the symmetrical patterns that are also shown. These patterns are known as supramolecular synthons and they are described in greater detail in Section 3.3. These length and angle parameters are obtained from around 500 and 150 accurately determined crystal structures, respectively. The geometrical parameters of the strong O–H···O bonds cluster within a narrow d-θ region. In other words, the distance and angle spread among the 500 crystal structures is not large. Therefore, one can obtain essentially correct information on the geometry of an O–H···O hydrogen bond from *any* structure that contains such an interaction. Of course, the crystal structure selected would need to be accurately determined but beyond that, it would matter little as to which particular structure was chosen. For the C–H···O bonds, however, this is not the case. The illustration shows that these interactions have lengths and angles that vary over relatively wide ranges. This happens because they are weak and therefore easily deformed by other interactions. Accordingly, estimates of the "normal" length and angle of a C–H···O hydrogen bond would vary widely and in fact, no single structure determination would enable such quantities to be reasonably determined. However, the d-θ scatter plot for the C–H···O bonds shows an overall directionality behaviour that is characteristic of all hydrogen bonds. So, we conclude that the C–H···O interaction is a hydrogen bond, but it is "soft" and easily deformed.

The CSD may be used to study not only individual interactions but also patterns of interactions. Individual hydrogen bonds and other intermolecular interactions are in themselves weak. A number of such weak interactions form more stable structural patterns that are called supramolecular synthons, as has been mentioned in the section above. The identification of robust, in other words, frequently occurring, synthons is a profitable exercise in crystal engineering. Robust synthons are more useful in crystal design because they have a higher likelihood of occurring in a new crystal structure. Robustness of supramolecular synthons is established by finding out the probability of occurrence of common structural patterns constructed with O–H···O, O–H···N, N–H···O and N–H···N hydrogen bonds. This data can be retrieved from the CSD. Database analysis shows that certain synthons like the carboxylic acid–pyridine pattern synthon constructed with O–H···N and C–H···O hydrogen bonds is a recognition motif between unlike functional groups that has a success rate of over 90% in crystals. Such analyses provide a strong argument in favor of synthesizing tailored molecules with functional groups in appropriate positions on the molecular scaffold for systematic crystal design. This is detailed in Chapter 3.

Database studies have been used to trace similarities between the directionality and interaction patterns of O–H···O hydrogen bonds and secondary bonds, namely those involving heavy p-block elements such as Bi(III) and chloride ions. It is seen, for example, that the ring and chain networks of Cl–Bi···Cl bonds in BiR_2Cl type crystal structures are topologically identical to O–H···O hydrogen bond networks in crystalline monoalcohols.

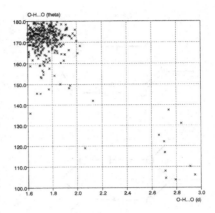

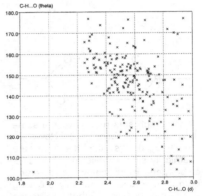

Scatter plots for O–H···O bonds in carboxylic acid dimers (top) and for C–H···O bonds in selected α,β-unsaturated carbonyl compounds (bottom). Notice the much tighter distribution for the stronger O–H···O hydrogen bonds.

All the graph sets here have the same designation, namely, R_2^2 (8). However, the two motifs on the top are chemically similar while the boric acid-carboxylate motif below is different: both the hydrogen bond donors belong to one molecule and both the hydrogen bond acceptors belong to another molecule.

2.6.2.1   *Graph Sets*

It is useful to have a scheme of describing the networks of hydrogen bonds in a crystal, in other words their connectivity and topology. Chemically different structures can have similar topologies and this may be important in crystal design where targets are often defined in terms of networks and the connectivity of nodes in these nodes.

Such a system of nomenclature of networks was suggested by Margaret Etter and applied to hydrogen bonds. Using graph set analysis, hydrogen bonds in crystals may be analysed in terms of four basic patterns. These patterns are chains (**C**), rings (**R**), intramolecular patterns (**S**) and other finite patterns (**D**). To these basic designators are added a subscript (**d**) and superscript (**a**) that denote the numbers of donors and acceptors in the pattern.

Graph set notations have been defined for simple and complex systems of hydrogen bonded patterns. In the simple patterns, there is just one type of symmetry independent hydrogen bond. These patterns are called motifs. In more complex cases, there is more than one motif in a structure. Graph sets can be computationally derived for a structure using the software associated with the CSD. The graph set notation is a purely descriptive exercise and provides a kind of structural taxonomy based on hydrogen bond topology. It helps in the identification of structural similarity between different crystal structures.

2.6.3   *Spectroscopy*

Vibrational spectroscopy is the classical method for the study of intermolecular interactions in condensed phases. The probes here are the vibrational frequencies of the atomic groups involved in an interaction, say a hydrogen bond, and because these frequencies can be measured very accurately, very subtle effects can be detected. The stretching frequency of the X–H bond in the IR spectrum will be lowered (bathochromic shift) when it is hydrogen bonded, say as X–H···Y–Z. These shifts can be relatively large (around 200 cm^{-1}) for strong hydrogen bonds or much smaller (around 20 cm^{-1}) for weak hydrogen bonds. Correlations between spectral parameters and hydrogen bond energies are also possible.

Despite the many benefits, the spectroscopic method is not free from drawbacks. Even for relatively simple systems, spectral complexity can prevent proper interpretation; this is the case in particular for systems exhibiting vibrational coupling. In consequence, the weakly activated C–H donor types are quite difficult to study with IR spectroscopic methods. This is also a serious problem for interactions like the halogen bond.

It may be advantageous to compare the spectroscopic properties in solids with other phases. In gas-phase rotational spectroscopy (microwave spectroscopy), the geometrical parameters of gas phase adducts are determined. These adducts can be observed in the ground vibrational state, free of interference from other molecules. This allows one to gain information on hydrogen bonds in the "pure" or undisturbed form. Apart from the geometries, dissociation energies, force constants and other parameters may also be derived. This kind of information is of fundamental importance because it is not obtainable from experiments in condensed phases where the equilibrium geometry is not directly observable.

IR spectroscopy is conveniently carried out in solution, and it is therefore an appropriate technique to use in the study of molecular aggregation in solution. Tetrolic acid, $CH_3C{\equiv}CCO_2H$, exists as hydrogen bonded dimers or as hydrogen bonded open chains, called catemers. Interestingly, the IR spectra of saturated solutions of this compound in ethanol and chloroform are different. These differences have to do with the nature of the molecular aggregation (dimer or catemer). From ethanol, we get a polymorph or crystal form of the compound that is based on the dimer pattern. From chloroform, we get another polymorph that is based on the catemer. This simple experiment, carried out by Roger Davey, shows that aggregates that exist in solution may carry over into the crystal. This result is important in elucidating the mechanism of crystallization — currently among the more challenging problems in crystal engineering.

2.6.4 *Computational Methods*

Quantum chemistry complements crystallography and spectroscopy in the study of intermolecular interactions. Theoretical methods can provide benchmark values for the energies of intermolecular interactions without the complicating effects of the solid state or solution environment. Crystallography and spectroscopy provide information on equilibrium geometries, as indeed most experimental techniques do. Computational methods on the other hand can be used to study domains of the potential energy surface which are far from the equilibrium structure. Such investigations are useful in the study of phenomena like crystallization which is a non-equilibrium event. Because of the widespread availability of powerful and yet low-cost computers, theoretical methods are now accessible to a large number of structural chemists, including experimentalists. Theory coupled with database research has emerged as an effective way of studying weak intermolecular interactions.

Rapid advances in the number crunching ability of computers also mean that computational results age much faster than experimental results. As an example,

Typical energy values for common intermolecular interactions (in kcal mol⁻¹)

van der Waals	0.5-2.0
Hydrogen bonds (quasi-covalent)	>20
Hydrogen bonds (strong)	4-20
Hydrogen bonds (weak)	1-4
Halogen bonds	1-20
Aurophilic	~10
Coordination bonds	~30-120

let us consider the fact that around a decade ago most computational work on intermolecular interactions was carried out using semi-empirical methods. These methods were state-of-the-art at that time but were just not designed or parameterized to treat intermolecular interactions properly. They were used nevertheless because there was no alternative. Today, it is doubtful if one would want to rigorously consider any of this older work, though general chemical trends might be still correctly predicted. Sometimes, even these trends are at variance with what is derived from present day *ab initio* methods. The latter are preferred today because they give reliable interaction energies in most cases to various degrees of approximation. Even for ab initio methods, however, the results are very dependent on the choice of basis set, the use of density functional theory (DFT) and the various ways in which electron correlation effects are handled. All these aspects and limitations must be considered carefully when computational work is assessed.

2.6.4.1   *Crystal Structure Prediction*

One of the most interesting applications of computational techniques in crystal engineering is called crystal structure prediction or CSP. Predicting a crystal structure computationally is very difficult. The problem of CSP is stated in the following manner: Given the structural formula of a small organic molecule with less than say 20 non-hydrogen atoms and with no more than two or three conformationally flexible C–C or C–X bonds, is it possible to predict its crystal structure with the kind of accuracy that is obtained with an X-ray diffractometer? What CSP demands is that the computational chemist predict correctly the unit cell, correct to about two decimal places, the space group, and the positional parameters of all the atoms in the asymmetric unit if what is given is only the structural formula of the compound. Research groups from all over the world have taken up this challenge and much of their effort has been streamlined in the form of blind tests that have been conducted, every few years, by the CCDC since 1999.

The need for CSP was highlighted in 1989 in Nature by the then editor John Maddox. He wrote that it was "one of the continuing scandals" that a general method for the prediction of crystal structures from molecular structures was not available. This provocative

Some molecules that have been given for CSP in the most recent blind test.

piece was much quoted in articles devoted to the then just evolving subject of crystal engineering, the design of organic solids with specific and desired properties. Twenty years later, much progress has been made. The main technique is to develop better atom potentials in potential energy expressions (like the Lennard-Jones expression mentioned in Section 2.2.1) and to use these better potentials to minimize the energy in a hypothetical crystal structure.

Correct structures will be predicted if the atom potentials used are accurate enough. If the most stable crystal structure, also called the thermodynamic structure, is not the same as the structure that is obtained in the experiment, the result of the CSP is in doubt. In this case, the experimental structure is known as the kinetic structure, and has a higher energy. This conflict between the kinetic and thermodynamic structure is a recurring problem in chemistry in general and in particular it is an important issue in crystal engineering. This is discussed in greater detail in Section 4.3.

2.7 Analysis of Typical Crystal Structures

A very important component in the subject of crystal engineering is the analysis of crystal structures. How do we understand a crystal structure as a combination of different kinds of intermolecular interactions? It is convenient to dissect a crystal packing and only look at a few molecules at a time so that only or a few interactions are highlighted. By looking at several sets or clusters of molecules, we obtain different views of the packing. In the end, we combine all these views and get an overall perspective of the crystal structure.

We will consider the crystal structures of five typical examples. To start with, let us consider the archetypical planar aromatic hydrocarbon, naphthalene, $C_{10}H_8$. Here, the aromatic rings pack with $C{\cdots}H$ interactions in what is called the herringbone arrangement, because of its resemblance to the backbone of a fish. In the herringbone packing, adjacent rings are inclined at angles of 50–90° with the hydrogen atoms of one ring pointing toward the carbon atoms of the other. Naphthalene adopts such a structure, with neighboring rings inclined at around 50° to each other. The molecule contains only C and H atoms in nearly equal stoichiometries and the predominant driving force in

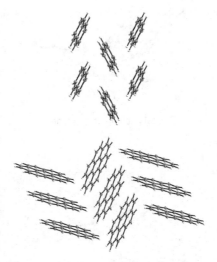

Crystal packing of naphthalene (top) and coronene (bottom).

adopting this packing is to maximise the number of C···H interactions. Naphthalene is a very important structure in the study of molecular crystals. It is as important in crystal engineering as is the structure of rock salt in inorganic solid state chemistry.

Coronene, $C_{24}H_{12}$, is a nice contrast to naphthalene. Coronene is also a condensed aromatic hydrocarbon but the C:H stoichiometry is rather different from what it is in naphthalene. The large carbonaceous core of the molecule tends to lead the packing towards C···C stacking interactions of the graphite type. These are also called π···π interactions. The molecules are stacked with a crystallographic short axis repeat of 4.7 Å. However, the peripheral H-atoms also favor the adoption of C···H interactions of the herringbone type. The final packing consists of both herringbone and stacking interactions. This is possible if the interplanar angles become large, around 80–90°. A schematic of the coronene packing is seen in the very common brick arrangement shown here.

We will next consider acridone, another aromatic molecule, which contains two hetero atoms. Here, different multiple interactions dominate the packing. N–H···O interactions assemble the molecules into chains. Molecules in the adjacent chains are stacked through π···π interactions in a nearly perpendicular direction. In addition, weaker herringbone interactions connect the chains in the third dimension. The crystal packing of acridone shows a similarity to the packing of the related molecules, 9,10-anthraquinone and indigo. The acridone structure resembles anthraquinone in its herringbone interactions and indigo in its hydrogen bonds. We note that it is easier to anticipate the strong hydrogen bonds in this structure but somewhat more difficult to predict the herringbone and π···π stacking interactions.

Our fourth example shows the triclinic form of quinhydrone which is the 1:1 molecular complex or co-crystal (see Section 6.6) of hydroquinone and 1,4-benzoquinone. Three kinds of intermolecular interactions make up the crystal structure. The first and strongest interaction is the O–H···O hydrogen bonds formed between the phenolic OH groups which act as donors and the carbonyl O-atoms which act as acceptors. These hydrogen bonds assemble the molecules into chains, which contain alternating molecules of hydroquinone and benzoquinone. In the second

This very common arrangement of bricks on a pavement is reminiscent of the crystal structure of coronene.

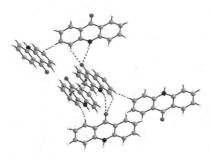

Crystal packing of acridone.

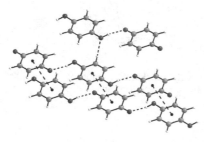

Crystal packing in 1:1 co-crystal of hydroquinone and 1,4-benzoquinone.

dimension, adjacent chains are stacked one over the other through $\pi\cdots\pi$ interactions. These interactions optimize the contacts between the electron poor quinone rings and the electron rich phenol rings. So effective are these charge transfer interactions that the color of quinhydrone crystals is blue black, in contrast to benzoquinone which is colored yellow and hydroquinone which is colorless. The final and weakest set of interactions in the quinhydrone structure are the weak C–H$\cdots$O hydrogen bonds formed by the quinonoid C–H groups and the phenolic O-atoms.

We notice that in many of these cases, there are three clearly distinct and separately identifiable sets of interactions in three nearly perpendicular orientations. The combined action of these three sets of orthogonal interactions results in the final, stable crystal packing.

Our fifth and final structure illustrates packing concepts in a coordination polymer. These compounds are described in detail in Chapter 7. The compound [Ag$_2$(4,4'-bpy)$_2$](BF$_4$)$_2$·2H$_2$O is a typical example of a linear coordination polymer. In this crystal, Ag(I) is coordinated to two 4,4'-bipyridine ligands forming an infinite chain. Two adjacent chains further interact strongly through $\pi\cdots\pi$ interactions between the pyridyl groups (3.3–4.2 Å). This arrangement is like a ladder structure. The shorter Ag$\cdots$Ag distances and the geometry of silver suggest the presence of argentophilic interactions (Section 2.5). The occurrence of counter anion, BF$_4^-$ for the cationic polymeric chain is further stabilised by water molecules through O–H$\cdots$F hydrogen bonding.

A typical crystal structure of a one-dimensional coordination polymer, [Ag$_2$(4,4'-bpy)$_2$](BF$_4$)$_2$·2H$_2$O

2.8 Summary

- Intermolecular interactions occur in the energy range of 1–20 kcal mol^{-1} and may be considered as the supramolecular "glue" that binds molecules in crystals.
- A crystal structure is the result of compromise between several weak intermolecular interactions.
- Strength, directionality and distance fall-off are three characteristic properties of intermolecular interactions.
- Interactions that are both strong and directional can be useful in crystal engineering.
- The close packing principle provides a geometrical approach to describe molecular crystals. In this approach, the shapes and sizes of molecules are important, and the interactions involved are isotropic or directionless. This model explains well the crystal structures of hydrocarbons.
- Anisotropic interactions result from hetero atoms like O, N and halogen. Such interactions cause deviations from close packing. Hydrogen bonding is the most significant anisotropic interaction in molecular crystals.

- A hydrogen bond is a stabilizing interaction between a donor X⋯H and acceptor Y⋯Z. Hydrogen bonds are directional. They have significant contributions from electrostatics, polarization and charge transfer (covalency) effects.
- Hydrogen bonds in crystals are hierarchic. Donor and acceptor strengths and the availability of donors and acceptors often determine the types of hydrogen bonds that occur in the crystal.
- The cooperative nature of hydrogen bonding is significant in organic crystals.
- Weak hydrogen bonds like C–H⋯O are electrostatic and can operate at distances longer than the van der Waals limit.
- Crystallography is the primary tool for studying intermolecular interactions. Spectroscopy and computational methods provide supporting evidence for quantifying the observations in the crystal.

2.9   Further Reading

Books

L. Pauling, *The Nature of the Chemical Bond*, 1939.
J. P. Glusker and K. N. Trueblood, *Crystal Structure Analysis: A Primer*, 1972, (third edition, 2010).
A. J. Pertsin and A. I. Kitaigorodskii, *The Atom-Atom Potential Method*, 1987.
T. C. W. Mak and G.-D. Zhou, *Crystallography in Modern Chemistry*, 1992.
G. R. Desiraju and T. Steiner, *The Weak Hydrogen Bond in Structural Chemistry and Biology*, 1999.
V. A. Parseghian, *Van der Waals Forces*, 2006.
G. Gilli and P. Gilli, *The Nature of the Hydrogen Bond*, 2009.

Papers

G. R. Desiraju, The C–H⋯O hydrogen bond in crystals. What is it? *Acc. Chem. Res.*, 24, 290–296, 1991.
M. C. Etter, Hydrogen bonds as design elements in organic chemistry, *J. Phys. Chem.*, 95, 4601–4610, 1991.
C. B. Aakeröy and K. R. Seddon, The hydrogen bond and crystal engineering, *Chem. Soc. Rev.*, 22, 397–407, 1993.
C. L. Schauer, E. Matwey, F. W. Fowler and J. W. Lauher, Silver coordination and hydrogen bonds: A study of competing forces, *Cryst. Eng.*, 1, 213–223, 1998.
J. D. Dunitz and A. Gavezzotti, Attractions and repulsions in molecular crystals: What can be learned from the crystal structures of condensed ring aromatic hydrocarbons? *Acc. Chem. Res.*, 32, 677–684, 1999.
G. R. Desiraju, Cryptic Crystallography, *Nature Mat.*, 1, 77–79, 2002.
C. V. K. Sharma, Crystal engineering — where do we go from here? *Cryst. Growth Des.*, 2, 465–474, 2002.
R. Paulini, K. Muller and F. Diederich, Orthogonal multipolar interactions in structural chemistry and biology, *Angew. Chem. Int. Ed.*, 44, 1788–1805, 2005.
P. Metrangolo, H. Neukirch, T. Pilati and G. Resnati, Halogen bonding based recognition processes: A world parallel to hydrogen bonding, *Acc. Chem. Res.*, 38, 386–395, 2005.
B. K. Saha, A. Nangia and M. Jaskolski, Crystal engineering with hydrogen bonds and halogen bonds, *CrystEngComm*, 7, 355–358, 2005.
J. D. Dunitz and A. Gavezzotti, Molecular recognition in organic crystals: Directed intermolecular bonds or nonlocalized bonding? *Angew. Chem. Int. Ed.*, 44, 1766–1787, 2005.
L. Brammer, G. M. Espallargas and S. Libri, Combining metals with halogen bonds, *CrystEngComm*, 10, 1712–1727, 2008.

S. L. Price, Computed crystal energy landscapes for understanding and predicting organic crystal structures and polymorphism, *Acc. Chem. Res.*, 42, 117–126, 2009.

C. Bissantz, B. Kuhn and M. Stahl, A medicinal chemist's guide to molecular interactions, *J. Med. Chem.*, 53, 5061–5084, 2010.

2.10 Problems

1. Most single component organic crystals have a packing coefficient (fraction of space occupied by molecules) between 65 and 77%. Why does the packing coefficient rarely exceed 75%?

2. What would happen if you try to crystallize a molecule that is so awkwardly shaped (for example, 1,4,5,8-tetra-*tert*-butylnaphthalene) that it cannot have a packing coefficient of even 65%?

3. A new definition for the hydrogen bond has been just accepted by the International Union of Pure and Applied Chemistry. Find this definition.

4. Hydrogen sulfide crystallizes at –60°C in a close packed structure with each molecule having 12 nearest neighbors. There is no hydrogen bonding. Why is this crystal structure different from that of ice?

5. How many crystal structures are there in the space groups $P2$ and $P2_1$ in the CSD? Give a reason for the large difference in these frequencies of occurrence.

6. Can you give a reason why crystals of proteins have a greater tendency to adopt high symmetry crystal systems (tetragonal, hexagonal, cubic) than do crystals of small organic molecules?

7. Most phenols exhibit O–H···O hydrogen bonding in the solid state. Very few show O–H···π bonding in these crystals. Give a reason for this observation.

8. The crystal structures of aminophenols have both O–H···N and N–H···O hydrogen bonds. Generally, which of these hydrogen bond types is the stronger? Why? Would aminoalcohols also be expected to show both these types of hydrogen bonding in their crystals?

9. Write down the graph set notations for the N–H···O hydrogen bond networks in benzamide, acetanilide and cyanuric acid.

10. Use the CSD to find typical distance ranges for the following interactions: O–H$_w$···O in water, N–H···O=C in 1° amides, Au···Au, O···Cl, C≡N···Br.

11. What is neutron diffraction? How is it different from X-ray diffraction? What are its uses in structural chemistry, especially in the study of hydrogen bonds?

Crystal Design Strategies

3

Crystal engineering is all about the design and synthesis of a crystal structure of a given molecule. Any such exercise involves the layout of a plan or blueprint, followed by the execution of the plan. Accordingly, the subject is subdivided conceptually into two parts, strategy and methodology. The basic methodology as to how crystal engineering is carried out has been described in Chapter 2, where we studied the nature and properties of important intermolecular interactions. In this chapter, we will evaluate the strategic component of crystal engineering, in other words, how the synthetic plan is worked out.

3.1 Synthesis in Chemistry

Synthesis is at the heart of chemistry. Chemistry is one of the few scientific subjects in which the researcher is free to make the object of his or her study. The chemist has the liberty to make molecules and compounds that are limited only by the fundamental laws of the subject and the imagination of the worker. Chemistry arose from alchemy and metallurgy, subjects which had to do with the preparation or extraction of new substances, and the transformations among these substances. Organic chemistry is said to have begun in 1828 with Wöhler's synthesis of urea from ammonium cyanate. With the chemist's increased knowledge of reactions, the complexity of synthetic targets increased. By the beginning of the 20th century, a molecule much bigger than urea, namely tropinone, was made by Willstätter in a complicated 17-step reaction sequence starting from the cyclic ketone suberone (cycloheptanone). By today's standards, tropinone is a small and rather simple molecule but 100 years ago, its synthesis was hailed as a masterpiece. Even today, however, a 17-step synthesis is difficult to conceive if one is thinking in terms

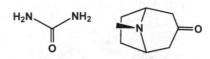

Urea (left) was the first organic molecule to be synthesized. This was in 1828. By 1906, a molecule that was as large as tropinone (right) was a challenging synthetic target.

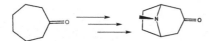

Willstätter (1906) made tropinone from suberone (left) in an elaborate 17-step reaction sequence.

Friedrich Wöhler (1800–1882).

Richard Willstätter (1872–1942).

of a strategy that builds up the molecule step-by-step in the forward direction. This is because such an exercise needs a complete mastery of organic reactions, and the ability to visualize the product right from the beginning of the reaction sequence. Such visualization is easy if the reaction sequence consists of a small number of steps but not if the sequence involves 17 steps. As textbook examples of easy reaction sequences in a synthesis, one can mention the single step preparation of methyl benzoate from benzoic acid by esterification or that of 4-nitroaniline from aniline in three steps by acetylation to acetanilide (protection), nitration to 4-nitroacetanilide followed by separation of the unwanted 2-nitroacetanilide and hydrolysis of the 4-nitro isomer to the desired product (deprotection).

Complex synthetic targets, on the other hand, do not lend themselves easily to such deductive strategies. As mentioned above, the Willstätter tropinone synthesis was already at the limit of this kind of approach, which is based on thinking about the product starting with the reactant. The same molecule was addressed by Robinson in 1917 with a radically different, inductive approach. Robinson made the target molecule *in a single step*, from succinic dialdehyde, methylamine and acetone. His reaction, shown here, illustrates the strategy. The synthetic thinking begins with the product rather than with the reactant, and the chemist selects reactions that will generate the product. Rather than working forwards from the reactant to the product, the chemist works backwards from the product to the reactant. Therefore, this strategy is called *retrosynthesis*. The contrast between Willstätter's 17-step sequence that uses a step-by-step approach and Robinson's single step synthesis that employs a retrosynthetic strategy is all too apparent.

What the chemist does in retrosynthesis is to analyze the product in terms of the bond connections that it contains. Some of these connections can be created through known reactions. Taking this process in reverse, one can *disconnect* a target molecule (conceptually) into smaller molecules that will react to form the product. This strategy of breaking down a target into smaller units is at the core of retrosynthetic thinking. After breaking down the target, in a paper disconnection, one carries out experiments on the smaller units so derived and creates the target through synthesis. Retrosynthesis is therefore nothing other than analysis, and the combined use of analysis and synthesis takes the chemist forward in making complex organic molecules.

Retrosynthesis is a remarkably powerful technique in organic synthesis. It was established and formalized in the 1960s by Elias J. Corey, who suggested the name *synthon* for the smaller sub-structural units that are generated by disconnecting the targets. The synthon is small but representative of the target. It contains those combinations of bonds in the target that can be made through known reactions. As the complexity of the targets increased, chemists found that new reactions had to be developed that could be used to make the more exotic synthons that these new targets entailed. It was sometimes preferable, for reasons of time and efficiency, to design a new reaction that could lead to a more complex synthon rather than to dissect the synthon further so that more standard reactions could be used. Corey's retrosynthetic approach is now the standard method for organic synthesis.

3.2 Supramolecular Chemistry

Supramolecular chemistry, or chemistry beyond the molecule, is the chemistry of the intermolecular bond and it is based on the theme of mutual recognition of molecules. Organic molecules recognize each other through a complex combination of geometrical and chemical factors and the complementary relationship between interacting molecules is characteristic of the recognition process. Molecules recognize one another through their dissimilarities rather than through their similarities. This was stated by noted chemists like Pauling. In a general philosophical context, it has been known since antiquity, and in different cultures all over the world, that recognition between dissimilar objects, that is those complementary in nature, may be more effective and efficient than recognition between similar objects. This complementary relationship between shapes (geometrical recognition) and between chemical functionalities (chemical recognition) is also what is important in supramolecular chemistry. Association between *different* molecular units is therefore what is relevant in assembly.

In 1806, Humphry Davy isolated crystals of chlorine hydrate and identified it as a loose addition compound of chlorine and water. Chlorine hydrate is probably the first multi-component crystal that was specifically characterized and the origins of supramolecular chemistry therefore go back at least 200 years. In 1948, H. M. Powell

Robert Robinson (1886–1975).

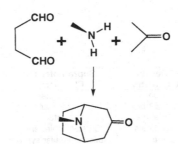

Robinson (1917) devised an ingenious single step synthesis tropinone based on retrosynthetic principles.

Things whose textures have such a mutual correspondence, that cavities fit solids, the cavities of the first the solids of the second, the cavities of the second the solids of the first, form the closest union.

Translated from Lucretius, *De Rerum Natura*

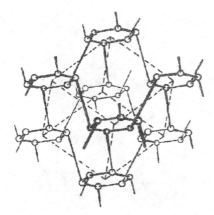

β-Hydroquinone network

Molecular Chemistry	Supramolecular Chemistry
Atom	Molecule
Covalent bond	Intermolecular bond
Molecule	Crystal
Synthesis	Crystal Engineering
Synthon	Supramolecular synthon
Isomer	Polymorph
Transition state	Nucleus
Reaction	Crystallization

This table gives the analogies between molecular and supramolecular chemistry. The first three rows are taken from the work of Lehn. The last five are from the crystal engineering literature.

described clathrates of hydroquinone in terms of their network structures. Describing a crystal structure as a network is a holistic and *non-reductionist approach* to structural chemistry. What is important is not the molecule itself but the topology of molecules as they arrange themselves in the crystal. Supramolecular chemistry, which stresses that the whole is more than the sum of the parts, is then all about the limitations of a reductionist approach to chemistry. *Reductionism* is the philosophy that maintains that the properties and behavior of larger entities can be expressed in terms of the corresponding attributes of the smaller components that make up the larger entity. Defining a crystal structure as a network is non-reductionist because we stress the "form" of the structure (larger entity) in topological terms and de-emphasize the molecule (smaller entity) itself. A manifestation of this holistic and non-reductionist argument is that different molecules can result in the same network.

Formal supramolecular thinking can be said to have begun with the work of Emil Fischer on the functioning of enzymes, way back in 1896. Fischer likened the working of an enzyme to the action of a key in a lock. The enzyme (lock) needs a small molecule (key) to make it work (open the lock). Any particular lock will open only with a particular key. Any particular key will only open a particular lock. A lock without a key is useless. A key without a lock is equally useless. This is called the lock-and-key principle. Supramolecular chemistry was further formalized through the work of Lehn on host-guest compounds. He stated that supermolecules are to molecules and the intermolecular bond what molecules are to atoms and to the covalent bond. Supramolecular chemistry as applied to the solid state brings one close to crystal engineering. In the same way that molecules are built from atoms with covalent bonds, crystals can be said to be built up with molecules using intermolecular interactions. J. D. Dunitz, in fact, stated that a molecular crystal is one of the best examples of a supramolecular entity. Accordingly, it can be argued that crystal engineering, which is the design of crystalline solids, is a supramolecular equivalent of organic synthesis. One can take the analogy further. A polymorph can most often be taken as the supramolecular equivalent of a structural isomer, while crystallization itself is nicely likened to a supramolecular reaction.

The molecular crystal is an example of a nearly perfect periodic self-assembly of millions of molecules held together by long range, non-covalent forces to produce matter of microscopic dimensions. So, crystals are supramolecular systems at an amazing level of precision. A high degree of order in a crystal structure is the result of complementary disposition of shapes, features and functional groups, in effect, geometrical and chemical recognition between neighboring molecules. From such ideas, it is not difficult to identify crystal engineering as a form of supramolecular synthesis.

3.3 The Synthon in Crystal Engineering

Strategic considerations may therefore be used in crystal engineering. In 1995, Gautam Desiraju proposed the idea of the *supramolecular synthon*, which he showed plays a role very similar to that played by Corey's synthon in organic synthesis. Central to his idea, is the act of defining an organic crystal structure as a network. This network consists, as all networks do, of nodes (molecules) and node connections (interactions). A very simple example is provided in the crystal structure of hydrogen cyanide, HCN. This crystal structure consists of a linear sequence of HCN molecules linked with $C-H\cdots N\equiv C$ hydrogen bonds. The node in this linear network is the HCN molecule and the node connection is the $H^{\delta+}\cdots N^{\delta-}$ interaction. Working back from the crystal structure, one may argue that any linear bifunctional molecule (containing a C–H group and a $C\equiv N$ group) that can form a $C-H\cdots N\equiv C$ interaction, will take the same network structure. Accordingly, one can understand the crystal structure of cyanoacetylene, $H-C\equiv C-C\equiv N$, which assembles with the same linear array as seen in HCN. One can go further, and introduce a phenyl ring between the hydrogen bonding functionalities. So, 4-ethynylcyanobenzene also adopts the same structure, as does the biphenyl analogue, 4′-ethynyl-4-cyanobiphenyl.

Important consequences follow even from simple examples like this. The node connections, in other words the hydrogen bonds, can be disconnected and one obtains the two relevant functional groups, C–H and $C\equiv N$. Any compound that contains these groups, preferably linearly situated in the molecule, is likely to adopt a linear hydrogen bonded structure. The $C-H\cdots N\equiv C$ hydrogen bond is called a *supramolecular synthon* and it may be dissected in the same way that a

Supramolecular synthons are structural units within supermolecules which can be formed and/or assembled by known or conceivable synthetic operations involving intermolecular interactions.

G. R. Desiraju, Angew. *Chem. Int. Ed.* 34, 2311–2327, 1995.

$C-H\cdots N$ hydrogen bonds are used, in exactly the same way, to build up the crystal structures of hydrogen cyanide, cyanoacetylene, 4-ethynyl-cyanobenzene and 4′-ethynyl-4-cyanobiphenyl.

Around the time Desiraju introduced the concept of the synthon, Wuest put forth the idea of a tecton, a molecule that has "sticky" regions that will associate with other molecules in the crystal. So, for example, adamantane-1,3,5,7-tetracarboxylic acid is a tecton with four connections. A tecton is essentially a node. In the end, a tecton is a molecule but there are some conceptual advantages in redefining some symmetrical molecules as tectons. Some authors have referred to the synthon–tecton model as a blueprint for organic crystal engineering. A tecton is a neutral or ionic species as opposed to a synthon which is a sub-structural unit. Therefore the concept of a tecton is particularly useful in the study of coordination polymers.

molecular synthon is dissected in organic synthesis. A supramolecular synthon was defined by Desiraju as a sub-structural unit in a molecular crystal that can be assembled with known or conceivable synthetic operations. This definition is based on Corey's definition of a synthon and the fact that it can be used effectively in crystal engineering shows once again that, conceptually speaking, crystal engineering is a supramolecular equivalent of organic synthesis.

A supramolecular synthon is a pattern that is composed of molecular and supramolecular elements. In a crystal, we can identify various patterns of interacting groups and functionalities. These patterns may repeat in other crystal structures composed of molecules that contain similar functional groups. For example, all molecules that contain a C–H group and a C≡N group can, in principle, form a C–H···N≡C hydrogen bond (see Section 3.3.5 for a discussion for the compound 1,3,5-tricyanobenzene). When crystal patterns repeat regularly, one can more confidently call the pattern of interactions a supramolecular synthon. This is because, a reliable synthon is one that can be disconnected to give a whole set of molecules that will, in all likelihood, assemble upon crystallization, to yield that particular synthon. So synthons that are made up of strong and/or directional interactions are possibly more useful in crystal engineering strategies. They are the ones that we would predict in experimental crystal structures. Accordingly, the synthons that are shown in the panel (next page) find frequent use in crystal design strategies.

3.3.1   *Some Representative Synthons*

The most commonly used synthons consist of a few functional groups held together by strong and fairly directional interactions. The panel (next page) gives a few typical examples. The carboxyl dimer synthon leads the crystal engineer to benzoic acid, while the catemers which is an infinite O–H···O patterns (Section 3.3.2) takes one to acetic acid. The catemer is favored in acetic acid because of the presence of the auxiliary, or supportive, C–H···O interaction that is possible for acetic acid but not for benzoic acid. This is discussed again in Section 3.3.2. Closely related to the carboxyl dimer synthon is the N–H···O based synthon formed by primary amides. However, because of the "extra" H-atom attached to the N-atom, the cyclic

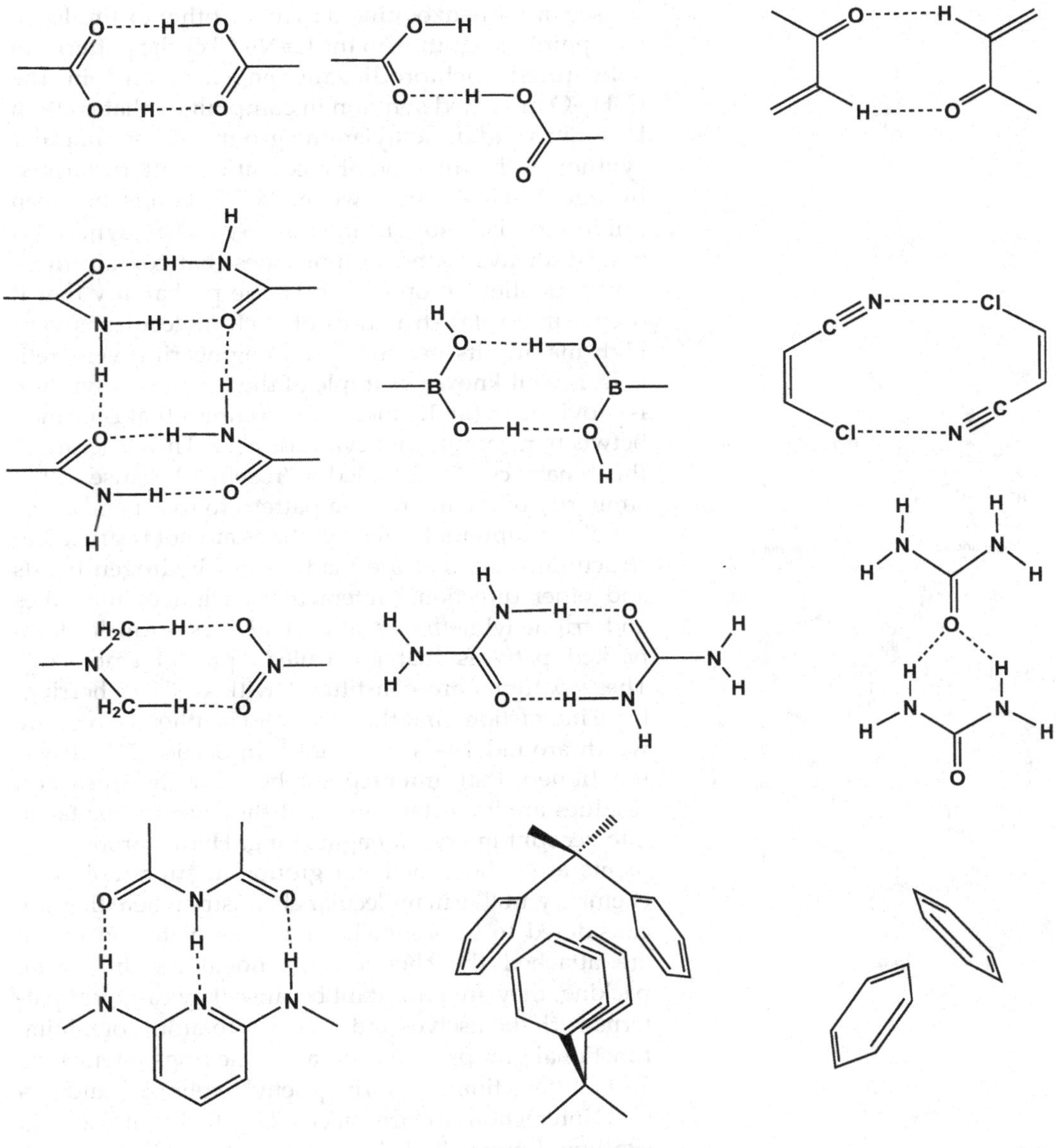

Some commonly used synthons.

synthon can be extended into a linear tape. Benzamide is a typical molecule that can be derived from the amide tape synthon. Such dimers are also seen in ureas and thioureas, where they occur as alternatives to a bifurcated pattern that is made up of translation related molecules. Somewhat related is the O–H group mediated tape found in boronic acids. A very common C–H···O based synthon is the symmetrical dimer that we see in 1,4-benzoquinone. We see other examples of two-point recognition in the C≡N···Cl cyclic patterns in substituted dichlorodicyanobenzenes, and in the C–H···O mediated synthon in compounds that contain both nitro and dimethylamino groups. A very popular synthon is the three point recognition pattern formed by one N–H···N and two N–H···O bonds between imide and diaminopyridine fragments. This synthon is almost always found in molecules that contain these functionalities. In other words, the probability that it occurs in crystal structures of such molecules is very high making its use in crystal engineering very reliable. A well-known example of the use of this synthon is provided in the 1:1 molecular complex that is formed between melamine and cyanuric acid. The structure of this binary crystal is called a "rosette" because of the similarity of the interaction pattern to rose petals.

Robust supramolecular synthons are not restricted to structural units that are made up of hydrogen bonds and other directional interactions. Diphenylmethanes and triphenylmethanes almost always furnish close-packed patterns that are called "phenyl embraces". These patterns are constituted with C···H–C herringbone interactions and they are valid synthons which are worth around 10–14 kcal mol^{-1}. In Section 2.1, it was mentioned that interactions between hydrocarbon residues are important and that they need to be taken into account in crystal engineering. Hydrocarbon fragments are valid functional groups in supramolecular chemistry, unlike in molecular chemistry where they are considered to be somewhat inert templates on which are attached the chemical functionalities. In crystal packing, they are important because they can form patterns with themselves and with hetero-atom containing functional groups. Typically, aromatic rings interact via C···H interactions to form phenyl embraces and via C···C interactions to form stacks. The stack is also a valid synthon. Linear alkyl chains longer than C$_4$ close pack in a hexagonal fashion and can be likened to a bundle of pencils. This type of close-packing is called a hydrophobic effect and these C···H interactions between alkyl groups are also justifiable synthons.

3.3.2   *The Carboxyl Dimer Synthon*

Let us consider now the very common carboxyl dimer, which has probably been studied in more detail than

Cyanuric acid   Melamine

Rosette

any other synthon. Benzoic acid exists in the form of hydrogen bonded dimers in the gaseous, liquid and solid states. In this crystal structure, the acid dimer is the operative synthon. The crystal as a whole is obtained by close packing of acid dimers. Using benzoic acid as a starting point it is easy to understand why terephthalic acid (benzene-1,4-dicarboxylic acid) forms a linear arrangement, or tape, of molecules in the crystal. The tape is one dimensional because it is constructed with dimer synthons on either side of the aromatic ring. The benzoic acid isolated dimer is a zero dimensional entity. Continuing in this series, we can understand why trimesic acid (benzene-1,3,5-tricarboxylic acid) forms a

(a)

(b)

(c)   (d)

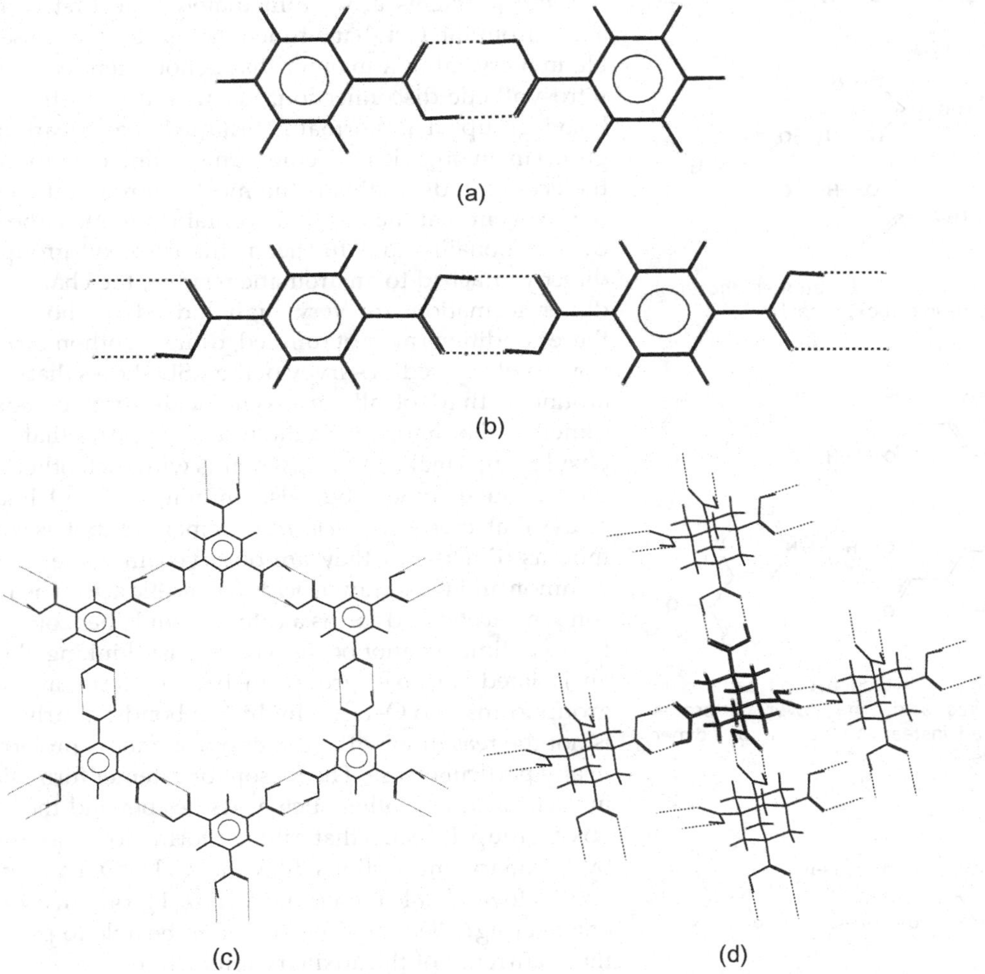

Building up the dimensionality of a crystal structure with the same synthon: (a) benzoic acid, (b) terephthalic acid, (c) trimesic acid and (d) adamantane-1,3,5,7-tetracarboxylic acid.

planar chicken-wire network of molecules in which each acid molecule is linked to three others through the dimer synthons, which we have already seen in benzoic acid and terephthalic acid. The trimesic acid network is an example of a two dimensional pattern. Finally, adamantane-1,3,5,7-tetracarboxylic acid provides an example of a three-dimensional network that is assembled with the same dimer synthon. Each adamantane ring sits at a tetrahedral node of the structure and the hydrogen bonded dimers act as node connectors. In all these structures, the carboxyl groups assemble during crystallization to form the dimer synthon. The final topology of molecules depends on the location of carboxyl groups in the molecule.

Some problems arise immediately. The first comes about from the fact that a functional group may assemble in a crystal in a manner that is not intended in the retrosynthetic disconnection. Let us return to the carboxyl group. It is normally assumed that a carboxyl group in an organic molecule forms a dimer synthon in the crystal. Indeed, this is the most common outcome. In the event that the molecule contains no other chemical functionality, and further if the carboxyl group is directly attached to an aromatic residue, the chances of dimer formation are very high indeed. If, however, these conditions are not fulfilled, dimer synthon formation is not assured. A survey of the CSD shows that only around a third of all carboxylic acids form carboxyl dimers. What happens to the rest? It appears that carboxyl groups make hydrogen bonds with each other not only to give dimers but also infinite O–H···O linear arrays that are called *catemers*. Catemers are not as common as dimers but they are observed in some rather common and well-known acids like acetic acid. The reason why acetic acid forms a catemer while benzoic acid forms a dimer cannot be derived by just thinking about an isolated carboxyl group. In both patterns an acid group forms two O–H···O hydrogen bonds; clearly this is not the reason why the dimer or catemer is preferred in any particular case. The reason for catemer formation in acetic acid and other such acids lies beyond the carboxyl group. It seems that catemer formation is assisted by auxiliary interactions like the C–H···O hydrogen bond. How reliable then is the carboxyl group in crystal engineering? We would need then to be able to predict the occurrence of the auxiliary interaction.

Another factor that prevents the formation of a carboxyl dimer synthon is the presence of other hydrogen

Dimer (top) and catemer (bottom) patterns in carboxylic acids.

Alternative hydrogen bond patterns in nicotinic acid. The lower pattern is formed instead of the carboxyl dimer above.

Solvent interruption of carboxyl dimer synthon (idealized).

bond donors and acceptors in the molecule. What happens, for example, if the molecule contains, in addition to the carboxyl group, a hydroxyl group (hydrogen bond donor) or a nitro group or heterocyclic atom (hydrogen bond acceptors)? Some simple examples will illustrate the difficulties involved in this problem. In 4-hydroxybenzoic acid, the carboxyl dimer is formed and the phenolic groups are hydrogen bonded to one other. This is a very good example of *structural insulation*. Insulation is a much desired attribute in crystal engineering because it lends an element of predictability to the anticipation of new crystal structures. Similarly, in the nitrobenzoic acids, the nitro groups do not generally hydrogen bond to the carboxyl groups. However, this is not the case in compounds that contain both heterocyclic N-atoms and carboxylic acid functionalities. Here, O–H···N hydrogen bond formation between the acid group and the heterocyclic N-atom occurs around 90% of the time. Carboxyl dimer formation occurs only in 10% of the cases. In such a case, the N-atom is said to *interfere* (rather successfully) with dimer synthon formation. A representative example is furnished by pyridine-3-carboxylic acid (nicotinic acid). Unlike insulation which is desirable, structural interference is a problem that needs to be overcome in crystal engineering. This is done by determining a large number of crystal structures of compounds that contain the pertinent functionalities. Studies of many pyridine carboxylic acids have provided guidelines that help the crystal engineer to predict when one of these acids will form a carboxyl dimer synthon and when it will form an O–H···N hydrogen bond.

Finally, some acids may form hydrates and solvates (with alcohols) in which the solvents interrupt the hydrogen bonding in the dimer synthon. Such an occurrence is difficult to predict and an anticipation of solvation, or hydration, is an important research topic today in crystal engineering. This is discussed further in Section 6.4 All in all, however, the carboxyl dimer synthon is termed to be reasonably *robust* in that it may be generally expected to be seen in the crystal structure of many carboxylic acids.

3.3.3 *Structural Insulation in Crystal Engineering*

Let us continue with the O–H···O hydrogen bonded carboxyl dimer synthon. This particular synthon is

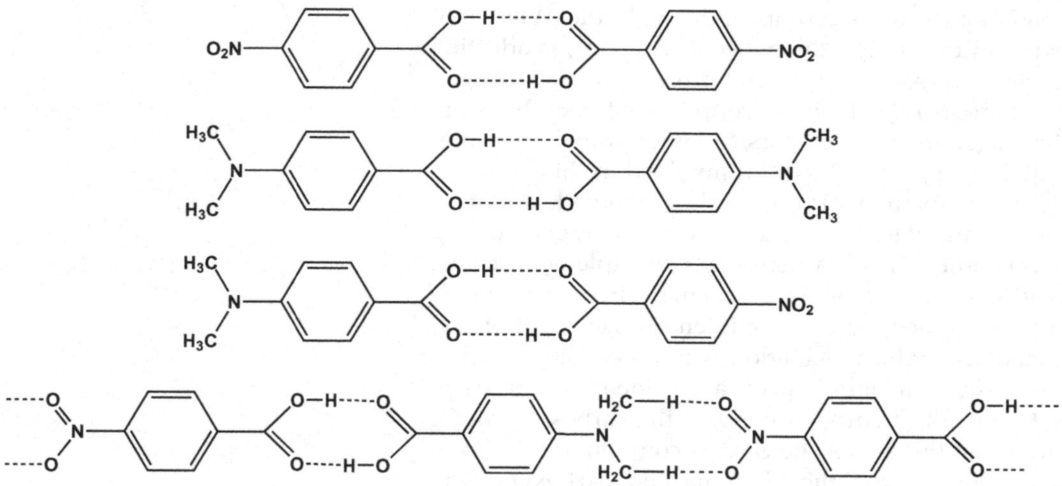

The 1:1 binary crystal of 4-nitrobenzoic acid and 4-(*N,N*-dimethylamino)benzoic acid is constituted with acid heterodimers, rather than the homodimers that occur in the native structures of the two acids.

overwhelmingly common in substituted benzoic acids and is seen in both 4-nitrobenzoic acid and in 4-(*N,N*-dimethylamino)benzoic acid. In both cases, the crystal structures are constituted with close packing of symmetrical acid dimers. These are called *homodimers* because the two acid molecules that form the dimer are identical. We can now perform an experiment and co-crystallize a 1:1 mixture of the two acids. The product obtained is the 1:1 molecular complex of the two acids (also called co-crystal in modern times). In the crystal structure of this binary complex the acid dimer is constituted with O–H···O hydrogen bonding between dissimilar acid molecules. This is called a *heterodimer* and its formation is energetically favoured over the homodimer. It is preferable, in this case, to form two acid heterodimers rather than two homodimers. The carboxyl dimer synthon in both cases is the same. It is called a *homosynthon* because the two functional groups that constitute it are the same (Chapter 6) but if the entire molecule is considered, we get a *homodimer* for the pure acids and a *heterodimer* in the binary crystal.

More interestingly, the heterodimers that are formed are linked to one another through the formation of C–H···O hydrogen bonds between the nitro and dimethylamino groups. Such hydrogen bonding is not uncommon and is found in other compounds that contain these functional groups. We can accordingly define a nitro–dimethylamino C–H···O hydrogen bonded synthon. The crystal structure of the 1:1 complex between the two above mentioned acids may therefore be described as a linear array of molecules held together by cyclic synthons that are made with O–H⋯O and C–H⋯O hydrogen bonds respectively.

What is noteworthy in this crystal structure is that the O–H⋯O and C–H⋯O domains are independent of one another. The nitro group is a hydrogen bond acceptor. Indeed this is its role in the present crystal structure. Why is it that the nitro group does not accept a hydrogen bond from the carboxyl group? In other words, why does the nitro group not interfere with the formation of the carboxyl heterodimer synthon? Similarly, the dimethylamino group is a weak C–H donor. Why does it not donate a C–H⋯O bond, to say, the carboxyl C=O of the acid? These are matters for further study. For the present, the crystal engineer should feel relieved that the strong and weak hydrogen bonded synthons (acid heterodimer and the nitro–dimethylamino dimer) are distinct. The acid group behaves as it would in the crystal structure of a simple acid (without nitro and dimethylamino groups), while the nitro and dimethylamino groups behave as they would in the crystal structures of simple compounds that contain these groups (without an acid group). These chemical functionalities are therefore well *insulated* from each other in the crystal. They do not interfere with one another.

Retrosynthetic thinking is particularly useful in the design of multi-component molecular crystals, also referred to as *co-crystals*. This is discussed further in Section 6.6.

3.3.4 *Discovery of New Synthons*

In the same way that new reactions were discovered so that synthons could be accessed more easily in organic synthesis, one can develop new supramolecular synthons that will permit the architecture of new crystal structures. An interesting synthon in this regard is the iodo⋯nitro synthon which is composed of two bifurcated polarisation induced $I^{\delta+}\cdots O^{\delta-}$ interactions. Halogen atoms are generally polarised negatively but under suitable conditions, they can be positively polarised, in other words they have electrophilic character. This tendency is the greatest in the heaviest and therefore most polarisable halogen, iodine. The iodo⋯nitro synthon is reasonably robust and is worth around 5 kcal mol^{-1}. The I⋯O interaction in this synthon

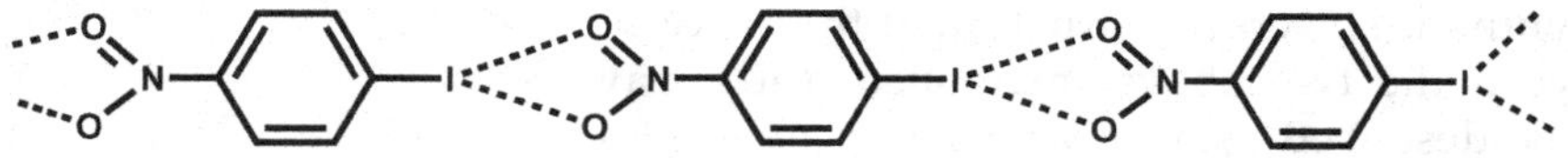

One component crystal

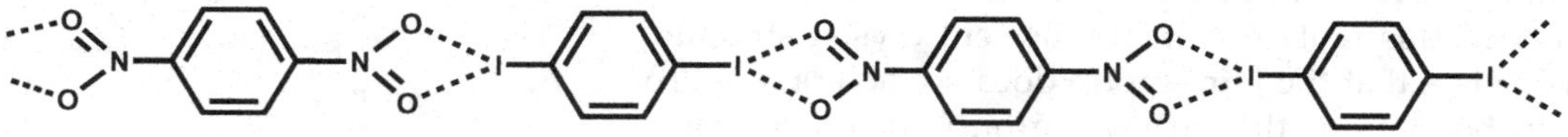

Two component crystal

Both crystals are built with the same synthon

can correctly be called a *halogen bond*. This synthon can be used in the same way as the carboxyl dimer synthon is used in terephthalic acid to build up linear arrays of molecules. So, one can understand the crystal structure of 4-iodonitrobenzene as a one- dimensional assembly of molecules linked with the iodo···nitro synthon. More interestingly, one can use the same synthon to generate a similar crystal structure of a multi-component crystal, the 1:1 molecular complex, or co-crystal, of 1,4-diiodobenzene and 1,4-dinitrobenzene. This result tells us that different molecular systems can have very similar crystal structures. In contrast, similar molecules can have very different crystal structures. Molecular and supramolecular chemistry are not the same and one cannot extrapolate too many supramolecular properties (like crystal structure) from molecular structures alone.

The iodo···nitro synthon is not exactly like the carboxyl dimer in its effectiveness as a linker. The formation of this synthon seems to depend much more on the location of iodo and nitro groups in a molecule than does the formation of the carboxyl dimer synthon on the location of carboxyl groups in a molecule. A simple example using *meta*- and *para*-substituted benzenes will illustrate this point. Terephthalic acid and 4-iodonitrobenzene have similar one-dimensional crystal structures. In both cases, the synthon connects phenyl rings to give a linear array of molecules.

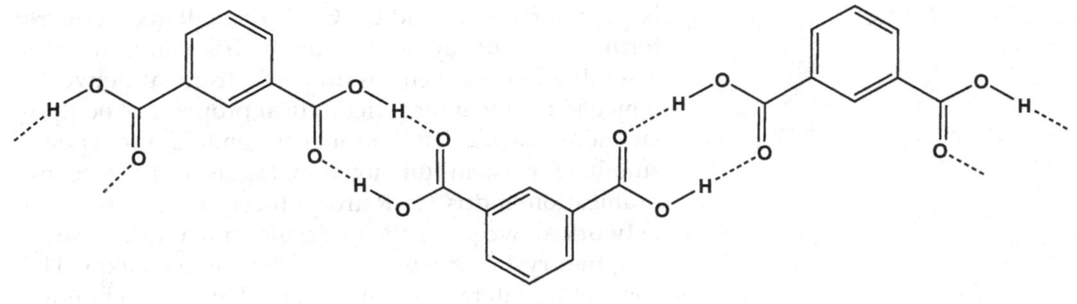

Zigzag tape in isophthalic acid.

But the behavior of the *meta*-substituted derivatives is in contrast to this. Isophthalic acid (benzene-1,3-dicarboxylic acid) forms a zigzag tape of molecules that are connected with the carboxyl dimer synthon. However, 3-iodonitrobenzene does not yield a topologically similar crystal structure. The reasons for this are unclear but the lesson for a crystal engineer is clear enough — synthons are not mere geometrical modules.

3.3.5 *Two-dimensional Patterns*

Retrosynthesis is a powerful technique in crystal engineering just as it is in molecular synthesis. We note that in the crystal structure of trimesic acid, which is a trisubstituted benzene of the type $C_6H_3X_3$, the molecule is a three connected node. The linkers are the acid dimer synthons are of the type $X\cdots X$. The topology of the resulting two-dimensional pattern is therefore a hexagonal chicken wire pattern. Knowing this, how would one generate a trigonal network? Retrosynthesis shows that we need a hexasubstituted benzene (or other hexagonally shaped molecule) of the type $C_6X_3Y_3$ in which the connecting interactions are of the type $X\cdots Y$. Referring back to C–H$\cdots$N hydrogen bonding, it is not difficult to identify 1,3,5-tricyanobenzene as a starting point in this crystal engineering exercise. Co-crystallization with hexamethylbenzene gives a layered 1:1 molecular complex. Alternating layers are composed of distinct molecules, either tricyanobenzene or hexamethylbenzene. The arrangement of tricyanobenzene molecules within layers is exactly as anticipated in the synthetic strategy.

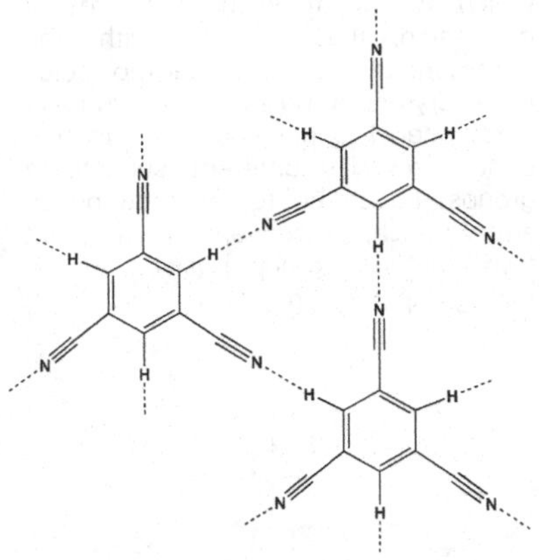

Other examples of trigonal layers formed by hexagonal shaped molecules are provided in the crystal structures of cyanuric chloride, $C_3N_3Cl_3$ where the connecting interactions are the polarization induced

Trigonal network in 1,3,5-tricyanobenzene constructed with C–H$\cdots$N hydrogen bonds.

N···Cl interactions, and the C–H···O mediated network formed by tribenzyl isocyanurate. Trigonal molecules assemble into noncentrosymmetric trigonal networks which show the interesting optical property of octupolar nonlinearity. This property renders the crystal suitable for use in functional materials in the telecommunication industry. Retrosynthesis of the trigonal network above gives the molecule tribenzyl isocyanurate, the crystal structure of which is shown below. The connecting interactions are C–H···O hydrogen bonds.

3.3.6   *Higher Dimensional Control*

Synthons, as defined above, define structural patterns in zero, one and two dimensions. These patterns can be combined or assembled, to yield higher dimensional arrays. Let us consider a simple one dimensional pattern such as the one formed by 4-iodo-4'-nitrobiphenyl. This pattern is structurally polar in that it has a directional sense. One can represent this sense of the array in terms of an arrow. Linear arrays of this type can be assembled to form a two-dimensional sheet. There are two simple ways of carrying out this operation. In the first case, all the arrows point in the same direction and the sheet formed has an overall structural polarity. In the second case, the arrows point in opposite directions and the sheet is symmetrical. Polar sheets can again be stacked in a parallel and antiparallel manner, but only parallel stacking will lead to a fully polar three-dimensional arrangement.

The trigonal networks described in Section 3.3.5 are examples of a polar two-dimensional structure. These networks can be stacked to give the three-dimensional structure in one of two ways: parallel stacking to give a polar structure and anti-parallel stacking to give a centrosymmetric structure.

It is observed that both 4-iodo-4'-nitrobiphenyl and tribenzyl isocyanurate form polar crystal structures. The space groups are noncentrosymmetric and both crystals satisfy one of the important tests for noncentrosymmetry, namely the generation of radiation of frequency 2ω when irradiated with a radiation of frequency ω. This property is known as *second harmonic generation* and it is important in the fabrication of new devices for telecommunication applications. Materials like 4-iodo-4'-nitrobiphenyl and tribenzyl isocyanurate are examples of crystals that have been designed with a particular application in mind.

There are two types of polarity that one refers to in solid state chemistry. Electrical polarity is concerned with the presence and alignment of dipole moments in a solid. Structural polarity, which is the topic of the present discussion, has to do with the alignment of unsymmetrical molecules in the crystal. A molecule like 4-iodo-4'-nitrobiphenyl is unsymmetrical in that either end has different substituent groups. It can be represented by an arrow which, by convention, can point from the iodo group position to the nitro group position.

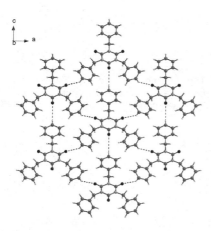

Crystal structure of tribenzyl isocyanurate.

Control of a crystal structure in higher dimensions is more difficult than making lower dimensional patterns. For example, the one dimensional iodo···nitro mediated pattern in 4-iodo-4'-nitrobiphenyl is easy to design. It is more difficult to predict whether or not this one dimensional pattern will form a polar sheet and finally a polar crystal. Likewise, one can design a trigonal sheet from $C_6X_3Y_3$ type molecules. However, it is much more difficult to reliably devise a strategy that will allow a parallel stacking of sheets, such as would be required for a polar three dimensional assembly which would give a second harmonic generation. This difficulty in achieving higher dimensional control arises from the fact that the interactions that are responsible for second and third dimensional assembly are usually weaker and less directional than those interactions that are used to constitute lower dimensional entities.

3.3.7 *Coordination Polymers as Networks*

The field of coordination polymers originated from the work of Robson who, in 1991, described the crystal structure of zinc cyanide in terms of a network consisting of tetrahedral zinc atoms as nodes and the cyanide ligands as connectors or linkers (node connectors). Each Zn-atom is connected to two C-atoms and to two N-atoms in a tetrahedral fashion. This crystal structure is topologically similar to that of diamond or cubic zinc sulfide. The term *coordination polymer* shows that the substance formed is both a co-ordination compound and also that it has an infinite structure. These compounds are described in detail in Chapter 7. In the context of crystal design strategies, Robson's work is one of the first examples of retrosynthetic thinking as applied to metal-organic compounds in the sense that the Zn···N connection may be likened to an organic supramolecular synthon. Assigning a network description to a coordination polymer is more instinctive than in a molecular organic solid because the metal atoms are easily identified as nodes. Applying a disconnection strategy to such a network results in the identification of metals and connecting ligands that are expected to yield a particular structure.

The identification and application of robust synthons, in other words, those which can direct crystal structures in a reproducible manner, is the main goal of synthetic strategies in crystal engineering. Generally,

Orpen synthon

supramolecular synthons are drawn from the field of organic chemistry and use interactions like hydrogen bonding for their assembly. The ability to use metal atoms in retrosynthetic strategies greatly increases the scope and ease of crystal synthesis. Guy Orpen, in 1998, showed that the metal assisted hydrogen bond affords a new class of synthon that permits the assembly of linear patterns. He argued that halogen atoms coordinatively attached to a metal ion can often act as strong hydrogen bond acceptors. Accordingly if the $[PtCl_4]^{2-}$ dianion is crystallized as a salt with an appropriate organic cation, such as 4,4'-bipyridinium dication, the result is a linear tape of molecules. The hydrogen bonds formed in such anion/cation combinations are strong because the donors and acceptors carry significant electrostatic charges. Therefore the synthons obtained thereby are expected to be robust.

3.3.8   *Useful Synthons*

Theoretically, the number of supramolecular synthons in a crystal structure is very large. The chemist is only limited by his or her imagination in the definition of a new synthon. The ultimate limit is the entire crystal structure in the same way that the ultimate Corey synthon for a molecule is the entire molecule itself. However, some synthons are more useful than others and it is important that the student becomes familiar with the reasons for labelling a synthon as useful.

(i)   A synthon is more useful if it occurs more frequently in a given group of compounds. This has to do with the strength of the intermolecular interactions that are involved, and their directionality. It also has to do with the commonness of the functional group and the ease with which it can be introduced into a molecule.

(ii)   A synthon is more useful if its formation is more specific. A carboxyl dimer synthon is more useful than an iodo···nitro synthon because the likelihood of its formation in an acid is greater than the likelihood of formation of an iodo···nitro synthon in a molecule that contains the iodo and nitro functional groups. We are referring here to the probability of the formation of a particular synthon in the crystal structure of a molecule that contains the requisite functional groups (with or without other chemical functionalities).

(iii) A synthon is more useful if it can lead to meaningful fragments in its disconnection. Consider, for example the C–H···H–C synthon. This is certainly a valid synthon and it occurs very commonly in crystal structures. However, retrosynthetic analysis of this synthon will not yield really useful fragments for use in crystal synthesis. All that one would learn is that a compound with a C–H group needs to be selected for crystallization! Synthons like Cl···N≡C or N–H···O or the carboxyl dimer synthon are much more useful.

(iv) The best synthons combine an economy in size with a surfeit of structural information. Small synthons are more common and hence they are likely to occur more frequently in new structures. However, they may not contain much specific structural information, as for example the O–H···O–H synthon in alcohols and phenols. Hence, they cannot be used so effectively in crystal structure planning. Large complex synthons contain plenty of structural information. However, they are so structurally limiting that they may not be observed very frequently. Somewhere in between, there is the golden mean of a reasonably small synthon that contains a maximum of useful structural information. These synthons are highly desirable in crystal engineering strategies. Many of these are shown in page 61.

(v) Larger synthons such as the melamine–cyanuric acid rosette can, however, be useful in the design of more subtle structural features. The examples given in this chapter mostly illustrate the building up of linear arrays. Planar and three-dimensional arrays have also been featured when the structure defining interactions are the same in all dimensions (1,3,5-tricyanobenzene, trimesic acid). A one-dimensional array is, however, not a complete crystal structure. Can one use synthons to build up two-dimensional structures and finally, the full three-dimensional structure when the structure defining interactions in the third dimension are not the same as that in the first and second dimension? The answer is yes, but as the synthons become larger and more three-dimensional, the interactions become feebler and less directional. These matters are the subject of current research in the subject.

3.4   Summary

- Crystal engineering is a form of solid state supramolecular synthesis.
- Retrosynthesis is the most effective way of approaching complex synthetic targets in both molecular and supramolecular chemistry.
- In crystal engineering, retrosynthesis is effected with the supramolecular synthon which is a sub-structural unit that contains molecular and supramolecular information.
- The advantage of using the synthon approach in crystal engineering is that it offers a simplification in the understanding of crystal structures. For example, the same carboxyl dimer is present in the structure of benzoic acid, terephthalic acid and trimesic acid; this carboxyl dimer can be used to systematically build up the dimensionality of the structure. Accordingly, the emphasis in crystal engineering is increasingly diverted from the constituent molecules to the topological features and geometrical connectivities of non-bonded interactions between the molecules.
- All crystal structures may be described as networks. The networks are constituted with nodes and node connections. It is profitable to describe structures in crystal engineering in terms of such networks. Retrosynthetic analysis may be performed accordingly on the network structure to yield the node structure (molecules) and the node connectivity information (supramolecular synthons).
- The advantages in using such an approach in crystal engineering are that: (i) molecule → supramolecule connections are easily established; (ii) comparisons between seemingly different crystal structures are facilitated; (iii) insulation is identified as a desirable attribute in a crystal structure and interference as a hindrance that needs to be overcome.

3.5   Further Reading

B. F. Hoskins and R. Robson, Design and construction of a new class of scaffolding-like materials comprising infinite polymeric frameworks of 3-D-linked molecular rods — a reappraisal of the $Zn(CN)_2$ and $Cd(CN)_2$ structures and the synthesis and structure of the diamond-related frameworks $[N(CH_3)_4][Cu^IZn^{II}(CN)_4]$ and $Cu^I[4,4',4'',4'''$-tetracyanotetraphenylmethane] $BF_4 \cdot xC_6H_5NO_2$, *J. Am. Chem. Soc.*, 112, 1546–1554, 1990.

M. Simard, D. Su and J. D. Wuest, Use of hydrogen bonds to control molecular aggregation — Self-assembly of 3-dimensional networks with large chambers, *J. Am. Chem. Soc.*, 113, 4696–4698, 1991.

D. S. Reddy, D. C. Craig, A. D. Rae and G. R. Desiraju, N···Br mediated diamondoid network in the crystalline complex carbon tetrabromide — hexamethylenetetramine, *J. Chem. Soc., Chem. Commun.*, 1737–1739, 1993.

G. R. Desiraju, Supramolecular synthons in crystal engineering — a new organic synthesis, *Angew. Chem., Int. Ed. Engl.*, 34, 2311–2327, 1995.

C. B. Aakeröy, Crystal engineering: Strategies and architectures, *Acta Cryst.*, B53, 569–586, 1997.

A. Nangia and G. R. Desiraju, Supramolecular synthons and pattern recognition, *Top. Curr. Chem. (Design of Organic Solids)*, 198, 57–95, 1998.

F. H. Allen, W. D. S. Motherwell, P. R. Raithby, G. P. Shields and R. Taylor, Systematic analysis of the probabilities of formation of bimolecular hydrogen bonded ring motifs in organic crystal structures, *New J. Chem.*, 23, 25–34, 1999.

J. Cai, C.-H. Chen, X.-L. Feng and X.-M. Chen, A novel supramolecular synthon for H-bonded coordination networks: Syntheses and structures of extended 2-dimensional cadmium(II) arenedisulfonates, *J. Chem. Soc., Dalton Trans.*, 2370–2375, 2001.

K. T. Holman, A. M. Pivovar, J. A. Swift and M. D. Ward, Metric engineering of soft molecular host frameworks, *Acc. Chem. Res.*, 34, 107–118, 2001.

R. Thaimattam, C. V. K. Sharma, A. Clearfield and G. R. Desiraju, Diamondoid and square grid networks in the same structure. Crystal engineering with the iodo···nitro supramolecular synthon, *Cryst. Growth Des.*, 1, 103–106, 2001.

C. B. Aakeröy, A. M. Beatty and B. A. Helfrich, A high yielding supramolecular reaction, *J. Am. Chem. Soc.*, 124, 14425–14432, 2002.

P. Vishweshwar, A. Nangia and V. M. Lynch, Recurrence of carboxylic acid-pyridine supramolecular synthon in the crystal structures of some pyrazinecarboxylic acids, *J. Org. Chem.*, 67, 556–565, 2002.

R. D. Willett, F. Awwadi, R. Butcher, S. Haddad and B. Twamley, The aryl bromine-halide ion synthon and its role in the control of the crystal structures of tetrahalocuprate(II) ions, *Cryst. Growth Des.*, 3, 301–311, 2003.

P. C. Crawford, A. L. Gillon, J. Green, A. G. Orpen, T. J. Podesta and S. V. Pritchard, Synthetic crystallography: Synthon mimicry and tecton elaboration in metallate anion salts, *CrystEngComm*, 6, 419–428, 2004.

J. J. Perry, G. J. McManus and M. J. Zaworotko, Sextuplet phenyl embrace in a metal-organic Kagomé lattice, *Chem. Commun.*, 2534–2535, 2004.

M. W. Hosseini, Molecular tectonics: From simple tectons to complex molecular networks, *Acc. Chem. Res.*, 38, 313–323, 2005.

C. B. Aakeröy J. Desper and B. Levin, Crystal engineering gone awry and the emergence of the boronic acid-carboxylate synthon, *CrystEngComm*, 7, 102–107, 2005.

C. B. Aakeröy, J. Desper and J. F. Urbina, Supramolecular reagents: Versatile tools for non-covalent synthesis, *Chem. Commun.*, 2820–2822, 2005.

M. Du, Z.-H. Zhang and X.-J. Zhao, Cocrystallization of bent dipyridyl type compounds with aromatic dicarboxylic acids: Effect of the geometries of building blocks on hydrogen-bonding supramolecular patterns, *Cryst. Growth Des.*, 5, 1199–1208, 2005.

L. S. Reddy, S. Basavoju, V. R. Vangala and A. Nangia, Hydrogen bonding in crystal structures of *N,N'-bis*(3-pyridyl)urea. Why is the N–H···O tape synthon absent in diaryl ureas with electron-withdrawing groups? *Cryst. Growth Des.*, 6, 161–173, 2006.

G. R. Desiraju, Crystal engineering: A holistic view, *Angew. Chem. Int. Ed.*, 46, 8342–8356, 2007.

F. C. Pigge, Losing control? Design of crystalline organic and metal-organic networks using conformationally flexible building blocks, *CrystEngComm*, 13, 1733–1748, 2011.

P. Sahoo, D. K. Kumar, S. R. Raghavan and P. Dastidar, Supramolecular synthons in designing low molecular mass gelling agents: L-amino acid methyl ester cinnamate salts and their anti-solvent-induced instant gelation, *Chem. Asian J.*, 6, 1038–1047, 2011.

3.6 Problems

1. Sketch possible crystal structure patterns for 3-cyanoethynylbenzene, given the experimental crystal structure of the 4-cyano isomer in this chapter. Assume that the C–H···N≡C interaction will be retained.

2. What is the maximum number of synthons in a crystal structure?

3. Sketch the crystal structure of the 1:1 molecular complex of CBr$_4$ and urotropin. You may assume that this structure is based on the N···Br interaction. Can you predict the crystal system for this binary compound?

4. The compounds 4-nitroethynylbenzene and 4-cyanoethynylbenzene have similar crystal structures. Can you suggest a reason for this?

5. The crystal structures of the 4-halobenzonitriles (hal = Cl, Br, I) consist of linear arrays of molecules joined with the hal...N synthon. All the structures are layered.

Sketch a diagram of a noncentrosymmetric crystal structure for this family, and three polymorphic centrosymmetric structures.

6. Sketch the trigonal sheet structure based on N–H$\cdots$O hydrogen bonding that one can expect for the symmetrical molecule cyanuric acid, $C_3H_3N_3O_3$. It is an unsolved problem today as to why this trigonal structure is not adopted by this compound. Check the CSD for the actual crystal structure of cyanuric acid.

7. Consider the centrosymmetric urea dimer shown in page 61. Use the CSD to find out the probability of occurrence of this dimer in all crystal structures that contain the urea molecule. Repeat this exercise for thiourea. Suggest a reason why the probability of occurrence of a symmetrical dimer is higher for thiourea than it is for urea.

8. Identify the supramolecular synthons in the DNA base pairs.

9. Ag(I) forms two different coordination polymers with 4,4'-bipyridyl. Charge neutralization is obtained with a non-coordinating anion in each case. These coordination polymers have a 1:1 and 1:2 stoichiometry. In both cases, all the N-atom sites in the ligand are coordinated by the metal atom. Given that Ag(I) can have variable coordination geometries (linear, trigonal, tetrahedral), sketch possible networks for both structures.

10. Consider the crystal structure of $Zn(CN)_2$ described in the text. You are required to maintain the diamond topology of this structure after changing the metal ion and the ligand. Suggest four coordination polymers that would satisfy these conditions.

Crystallization and Crystal Growth

4

Crystallization, or synonymously recrystallization, is fundamental to many natural, chemical and industrial processes. Crystallization is one of the oldest and most important unit operations, for example the extraction of salt crystals from sea water. The process occurs in a single step combining both purification and particle formation. Control over the crystallization process allows one to obtain products with desired and reproducible properties. The majority of commercial products based on fine chemicals, dyes, explosives, and photographic materials require a crystallization step during manufacture. Many drug products contain active pharmaceutical ingredients (API) and excipients in the crystalline state. The pharmaceutical industry in particular requires production of a desired crystal form (polymorph, co-crystal, solvate, hydrate) to assure bioavailability and stability of a drug substance. The quality of a crystalline product is usually judged on five main characteristics: chemical purity; crystallographic purity; morphology; size; size distribution.

If a crystal is an example of a supramolecular entity, crystallization is a supramolecular reaction. In this context, the properties of a crystal are defined by the intermolecular interactions that constitute it. Crystal engineering is the subject wherein the structural chemist simultaneously addresses structure, design and function. Understanding a solid state structure allows the crystal engineer to manipulate its chemistry so that one may make functional materials that incorporate features such as porosity, catalytically active groups and chirality. For industrially important solids, crystal engineering offers a valuable tool to tailor physicochemical properties of already marketed products. An in-depth knowledge of crystallization processes and solid state properties is therefore required to ensure consistent quality of crystal forms and to implement new and existing approaches for crystal engineering.

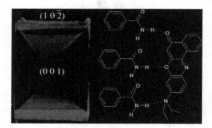

Crystallization is essentially a kinetic phenomenon. The symmetry of a crystal surface is lower than the bulk. Additives in the growing solution tend to get incorporated within evolving growth sectors that are terminated by faces that favor guest adsorption. During the growth of benzamide crystals from a solution containing the luminophore Nile red, the dye gets "painted" onto the {102} faces. The amount of dye that is adsorbed is so small that it cannot be detected with X-ray diffraction. Optical probes provide clues for recognizing the host-guest interaction at the molecular level.

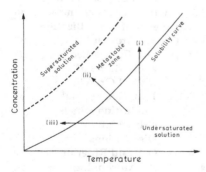

Solubility diagram for a solid.

Crystallization occurs only in the supersaturated region. A solution whose composition lies below the solubility curve is undersaturated. Crystallization from solution can be achieved by: (i) decreasing the solubility of the solid at constant temperature (isothermal evaporation); (ii) decreasing the vapor pressure of the solvent under reduced pressure and; (iii) cooling the solvent.

4.1   Crystallization of Organic Solids

Chemists invariably employ crystallization as a separation and purification technique. Crystallization refers to a first order phase change in which a solid product is isolated from a solution, melt, vapor or even from a solid phase. *Supersaturation* is a prerequisite for achieving crystallization from solution. For crystallization from a melt or liquid, *supercooling* is the analogous prerequisite. Even though liquid water freezes at 0°C at 1 atm, crystallization does not actually take place unless water is cooled to below 0°C. Crystallization occurs away from equilibrium and techniques of crystal growth for molecular solids are based on methods used to attain metastability of the mother phase.

4.1.1   *Solution Crystallization*

Solution crystallization refers to crystallization of an organic compound dissolved in a solvent or solvent mixture. It is normally conducted by dissolving the desired organic solid in a suitably "inert" solvent, one in which no chemical reaction occurs between solute and solvent. Solvent evaporation is the most common method adopted for increasing supersaturation levels, thereby inducing crystallization; evaporation can be performed at atmospheric or reduced pressure. A solubility diagram provides guidelines for the variation of concentration versus temperature in relation to saturation and solubility. The width of the metastable region depends on the presence of seeds and impurities. Once crystallization is initiated, the concentration in solution may follow different paths. The particular pathway taken may depend on the rate of evaporation, surface area of the crystals, secondary nucleation rate and inherent crystal growth rate. In addition, oiling out, agglomeration and uncontrolled nucleation leading to precipitation can occur if the concentration goes beyond the metastable region.

4.1.1.1   *Antisolvent Crystallization*

When a substance is highly soluble, *antisolvent* crystallization can be useful. In this method, another solvent (for example ethanol to water) is added, normally at ambient condition, to decrease the solubility of the desired substance. Pharmaceutical and fine chemical manufacturers frequently rely on antisolvent crystallization because heating and cooling of solutions is not involved. The technique is suitable for polymorph control and yield improvement. The solvent must be chosen such that it does not interact with the desired substance or form a co-crystal or a solvated crystal (Sections 6.3 and 6.4).

4.1.2 *Melt Crystallization*

Most organic solids have low melting points, generally less than 200°C. If a solid does not decompose on heating and melting, crystallization can be performed by melt cooling. If the solubility is sufficiently low, melt cooling is desirable for crystallization. Melt cooling is preferred when solution evaporation leads to solvent inclusion or co-crystal formation. The presence of impurities may cause eutectic formation.

4.1.3 *Sublimation*

Sublimation, or crystallization from the vapor, is a process in which a solid compound, when heated, directly passes into the vapor without going through the liquid state and produces crystals when the vapor passes over a cold surface. Every solid compound has a definite vapor pressure at a given temperature. Most organic solids exhibit low vapor pressures over a wide temperature range. Sublimation can be employed to grow a crystal provided the vapor pressure of the molecule of interest is high and the likely impurities have significantly lower vapor pressures. Relatively few organic compounds sublime at atmospheric pressure (1,4-benzoquinone, camphor, naphthalene, benzoic acid) because the vapor pressures of most organic compounds are too low. However, a reduction of the pressure will cause an increase in the sublimation rate. Notice that many organic solids that do sublime have characteristic smells.

Sublimation is the process of transforming a solid into vapor. Thus, sublimation enables the separation of a volatile organic molecule in a more pure form. By keeping the condenser at a temperature lower than the temperature of vaporization (supersaturation zone), crystals can be isolated. No artificial means of refrigeration are usually needed in sublimation, the air being sufficiently cool to condense the vapor into a solid form.

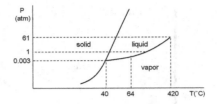

The phase diagram of a pure substance, say phenol, provides guidelines for growing crystals either by melt cooling or by sublimation. The equilibrium line, liquid-solid or vapor-solid respectively, gives the variation of melting point or sublimation point with pressure. Crystallization can occur either from supercooled liquid or supercondensed vapor.

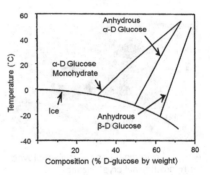

Phase diagram of a two-component system, water and α-D-glucose, at 1 atm. There are three eutectic points. Notice the variation of the solubility line with temperature for the hydrated and anhydrous forms. Crystallization of α-D-glucose hydrate is possible only below room temperature.

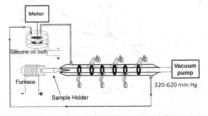

A simple device for crystallizing anhydrous organic solids based on gradient sublimation.

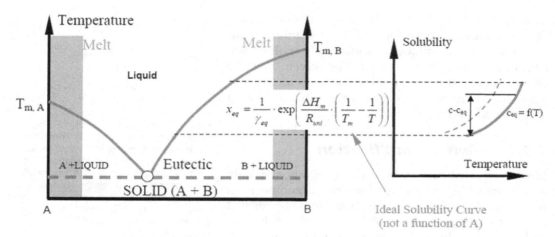

$$x_{eq} = \frac{1}{\gamma_{eq}} \cdot \exp\left(\frac{\Delta H_m}{R_{uni}} \cdot \left(\frac{1}{T_m} - \frac{1}{T}\right)\right)$$

Phase diagram for a two component mixture (A and B) at 1 atm. The eutectic composition limits the yield of the product prepared by melt crystallization. A comparison of melting point variation with temperature, and solubility versus temperature are shown here.

A typical acid digestion reactor employed for hydrothermal crystallization.

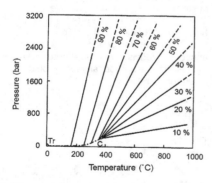

The diagram shows variation of the autogenous vapor pressure of water as a function of temperature in a hydrothermal vessel. The pressure also depends on the filling of the vessel (given in brackets).

4.1.4   *Hydrothermal and Solvothermal Crystallization*

The term solvothermal refers to a process wherein crystallization is carried out in the presence of a solvent at temperatures higher than the boiling point of the solvent. The reaction is carried out in a closed vessel called hydrothermal reactor or autoclave. Water is the most used solvent (hydrothermal) since it is a pressure transmitting medium. Increasing the temperature increases the vapour pressure (till critical temperature) above the aqueous solution. Increasing temperature has two effects: the viscosity and polarity (dielectric constant) of the solvent decreases. A lower viscosity favors higher mobility of the solute molecules. Increasing the temperature increases the solubility of a substance in many cases. Nature is a model for the crystallization of many minerals under hydrothermal conditions. The solvent selected should be one in which the solubility is sparing, but finite.

This method is well suited to coordination polymers. Generally the metal source and the organic ligands have different solubilities in the majority of the commonly available solvents including water. Hydrothermal/solvothermal condition with moderate temperatures (for water it is between 100°C and 250°C) are ideally suited for solubilising the reactants. In selected cases, a mineralizer (hydroxides, halides, alkali salts of weak acids, carbonic acid and boric acid) is used to increase the solubility.

4.1.5 *Crystallization from a Solid Phase*

In the previous sections we discussed crystallization of a solid phase from a melt, solution or vapor phase. A significant feature of the nucleation of a new compound from a solid phase is the importance of cooperativity in the crystal, unlike in the liquid or gaseous states. In solid state transformations each molecule is influenced by what happens in its vicinity. First-order transitions are generally believed to occur by nucleation of the new phase within the old phase followed by its growth. Nucleation is influenced by defects; depending on the nature of these defects, nuclei of the new phase are formed at different temperatures and grow at different rates.

Let us consider the crystallization of a new solid (daughter phase) from a parent solid (mother phase). Once the transformation is triggered it usually goes to completion. Yuri Mnyukh regarded this phase transformation as a non-equilibrium phenomenon like other crystallization processes; superheating or supercooling is also a necessary condition for solid state crystallization. Mnyukh considered solid state transition by a nucleation and growth mechanism; the nucleation step is critically dependent on the presence of suitable defects. According to him, the absence of orientational relationships between the parent and daughter phases is a general rule, but there are many exceptions.

4.1.5.1 *Single Crystal to Single Crystal (SCSC) Transformations*

A solid-state reaction, in which a reactant single crystal yields a product in the form of a single crystal, is commonly referred as a single-crystal-to-single-crystal (SCSC) reaction. A distinguishing feature that differentiates a SCSC reaction from other solid state reactions is that the single crystal maintains its integrity (retains shape and transparency) during the process. SCSC reactions open up alternative paths for reactions that are otherwise inaccessible, for example in the production of

Common solvents employed in solvothermal crystallization

Solvent	T_c (°C)	T_p (bar)
H_2O	374.1	221.2
CH_3OH	240	81
C_2H_5OH	243	63.8
DMF	377	43.6

T_c (°C) and Tp (bar) correspond to critical temperature and pressure.

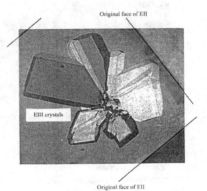

p-Dichlorobenzene EII to EIII heating transition

Original face of EII

EIII crystals

Original face of EII

Photomicrograph (partially polarized light) of an EII-α (monoclinic) to EIII-β (triclinic) transition in *p*-dichlorobenzene. The phase transition was initiated by a slight prick of a needle in the parent (EII-α which created nucleation sites for the growth of EIII-β in different directions.

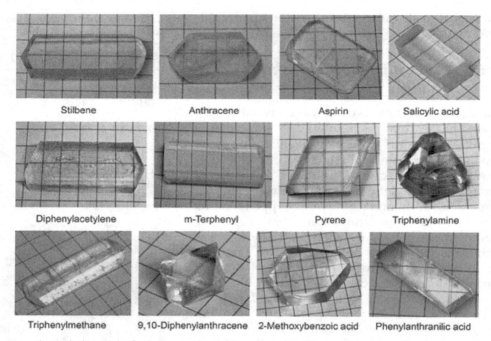

Large single crystals of substances grown by slow evaporation. The crystals are several mm in size.

The 230 space groups can be divided into centrosymmetric and non-centrosymmetric categories. Only the centrosymmetric space groups have inversion symmetry. Among the non-centrosymmetric space groups, there are 65 space groups that further lack reflection symmetry. Typical examples are $P1$, $P2_1$ and $P2_12_12_1$. Enantiopure substances can crystallize only in one of these 65 space groups. Leonard Sohnke (1874) was the first to derive these 65 space groups and hence these are referred to as Sohnke space groups. Achiral molecules may also crystallise in Sohnke space groups and in these cases the chirality of the crystal structure arises from chiral packing of the molecules (supramolecular chirality). A special subgroup of the Sohnke space groups are the 22 enantiomorphous space groups. If one enantiomer crystallises in an enantiomorphous space group, the opposite enantiomer will crystallise in the other space group of that pair. An example of an enantiomorphous pair is $P3_1$ and $P3_2$; the effect arises since equivalent points generated by a 3_1-axis will form a helix, and points generated by a 3_2-axis will form an enantiomorphous helix.

crystalline organic polymers. These reactions can also offer insight into solid state reaction mechanisms by monitoring crystal structure changes on the diffractometer.

4.1.5.2 *Mechanochemistry*

Mechanochemistry is a technique to grow crystals in a solvent free medium for crystals with poor solubilities under ambient condition. It involves grinding of solids in a pestle and mortar or ball mill. As early as 1844, Wöhler prepared the 1:1 quinone-hydroquinone molecular complex, quinhydrone, by a similar method. Recently the approach has been used for co-crystals and coordination polymers. However, the method suffers from the defect that good quality single crystals cannot be made. In dry grinding (also referred as neat grinding), the heat generated can induce local melting at the interface between different reactants; this causes the nucleation of a new phase. Sometimes, a small but controlled amount of solvent is required to enhance the rate of nucleation of the product phase (referred as kneading, liquid assisted grinding or solvent-drop grinding).

Anthranilic acid crystallizes as three polymorphs: in the commercially available Form I, two independent molecules occur in the asymmetric unit, one is unionized and the other is zwitterionic. Forms II and III have only unionized molecules. While Form II can be crystallized from different solvents, III is grown either by sublimation or from melt. These forms interconvert upon solvent drop or neat grinding.

A 1:1 co-crystal of caffeine and glutaric acid is obtained in two polymorphic modifications depending on the method of preparation. Neat grinding, or grinding in the presence of a nonpolar solvent such as *n*-hexane, results in Form I while kneading in the presence of a polar solvent such as chloroform favors Form II. The ternary co-crystal of 1,4-benzoquinone, anthracene and 2-naphthol may be obtained through neat grinding of the components.

4.1.6 *Crystallization of Chiral Solids*

A mixture of R and S enantiomers can crystallize in three different ways: as a racemic crystal, as a conglomerate of chiral crystals, or as a solid solution. A racemic crystal contains equal amounts of the two enantiomers associated at the molecular level. In a conglomerate, the two enantiomers crystallize as separate crystals. The whole collection contains an equal amount of R and S crystals. Under appropriate conditions (seeding), crystals of only one enantiomer may be obtained for compounds where the enantiomers can equilibrate freely (1,1'-binaphthol). This is called spontaneous resolution. Complete spontaneous resolution is used for absolute asymmetric synthesis of labile coordination compounds, organometallic reagents and chiral supramolecular materials. A conglomerate of equal amounts of R and S crystals melts at the eutectic temperature which is the lowest temperature at the solid-liquid interface in the phase diagram. In other words, this melting temperature is lower than the melting points of the pure R and pure S crystals. In a racemic crystal, however, the melting point may be higher or lower than the melting points of the pure R and S crystals. This is like the phase diagram of a 1:1 co-crystal and is akin to compound formation. Whether a substance crystallizes as a conglomerate or a racemate can be decided from the space group symmetry. Racemates

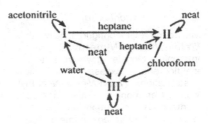

Interconversion of anthralinic acid polymorphs through mechanochemical routes.

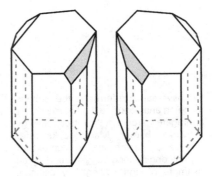

Pasteur originally held the opinion that chiral molecules could not be synthesised in the laboratory, not even as racemates. This was soon disproved by the synthesis of maleic and lactic acids followed by optical resolution using optically active bases. In 1894 Fisher reported the first asymmetric synthesis: transformation of hexoses to heptoses without the formation of diastereoisomers. In the late 1930's it

(Continued)

(Continued)

was found that configurationally labile N-allyl-N-ethyl-N-methyl-N-phenyl ammonium iodide gave rise to an enantiomeric excess on slow crystallization. N-allyl-N-ethyl-N-methyl-N-phenyl ammonium iodide undergoes spontaneous resolution on crystallization, *i.e.* the two enantiomers appear as separate crystals; since the salt is configurationally labile, the solution remains racemic during the crystallization and since all crystals occasionally grew from a single nucleus, an enantiomeric excess can be obtained. This was the first example of total spontaneous resolution. Total spontaneous resolution of prochiral reactive substrates or chiral chemical reagents has since been performed and such substances have been used in enantioselective synthesis, as well as in inter- and intramolecular photochemical reactions.

We have earlier referred to the nucleus as being analogous to a supramolecular aggregate around the free energy activation barrier. Many books refer this as a *cluster*. In the context of crystal engineering, aggregation is an example of enthalpy-entropy balance. The enthalpy effect is synthon based and is a measure of solvent-solute and solute-solute interactions. The entropy effect is associated with solvent released (positive entropy) or trapped (negative entropy) within the growing nucleus.

adopt space groups that contain inversion centres, mirror planes or glide planes. Enantiomers must crystallize in chiral space groups like $P1$, $P2_1$, $P2_12_12_1$, $P3_1$ and $P3_2$.

4.2   Nucleation

The primary goal of crystallization is to generate periodic solids with desired stoichiometry, chemical purity and shape in a reproducible manner because these individual characteristics can affect physical and chemical properties. Crystallization is a complex process that takes place in several stages: it is convenient to consider it occurring as two steps, nucleation and growth. Nucleation of a crystal from solution, melt or vapor is the first step that is achieved as soon as a nucleus with a particular aggregation of atoms, ions and/or molecules is stabilized from a system.

4.2.1   *Nucleation as Distinct from Crystal Growth*

Let us consider the more common example of a nucleation of a solute crystal from solution. As the system becomes supersaturated with solute, spontaneous appearance of a solid phase occurs. Supersaturation implies that the free energy of the initial solution phase is greater than the sum of the free energies of the solute crystals plus the final solution phase. A larger difference in free energies between these two terms provides greater driving for crystallization. The dissolved molecules begin to aggregate as the solution gets concentrated and this is explained in terms of nonbonding interactions between solute molecules (Chapter 3) which eventually form critical nuclei that act as centers for further growth.

Nucleation occurs either spontaneously or is induced artificially. Often, it is difficult to decide whether a system has nucleated on its own or if nucleation has been induced by an external stimulus. Nucleation may therefore be broadly considered in two categories: primary (homogenous) and secondary (heterogenous). In the former case the system is in a homogenous phase while in the latter nuclei occur in the vicinity of a second phase (seed, impurity, irregularity).

Nucleation is the process of generating crystal nuclei inside a large volume of the metastable solution phase; this transformation requires traversing a free energy barrier. This passage can be understood by considering the free energies associated with the formation of the *nucleus*. The difference between the free energy of the molecule of the bulk (ΔG_V) and that of the surface (ΔG_S) is referred as interfacial energy. This is always a positive term and acts to destabilise the nucleus. As a consequence, at very small sizes, below the critical size, many of the aggregating molecules will tend to reside at the surface making the nucleus unstable. The addition of even a single molecule will increase the free energy of the system. On average such a nucleus will tend to dissolve rather than grow. Once the nucleus gets sufficiently large, however, the decrease in free energy associated with formation of the bulk phase becomes high enough that the hindrance offered by the interfacial energy is not important. At this stage, addition of a molecule to the nucleus lowers the free energy. This intermediate size represents the activation barrier at which the free energy of the system is decreased whether the nucleus grows or dissolves; this is known as the *critical size*. The existence of a critical size has many implications. Nucleation is caused by concentration fluctuations that enable aggregation of molecules to exceed the critical size. The probability of nucleation strongly depends on the critical size which in turn is a function of the interfacial energy. As a consequence, nucleation can be manipulated by varying supersaturation or solution concentration. As supersaturation increases, both activation energy and critical size decrease. Consequently, the activation energy becomes so low that spontaneous and rapid nucleation occurs. The rate of nuclei formation is defined as the number of clusters that grow to greater than the critical size. Since activation energy is a decreasing function of supersaturation, the rate of nucleation increases with an increase in supersaturation. At larger nuclear radii the free energy of nucleation decreases and eventually becomes negative. The growth stage, which immediately follows nucleation, is governed by the rate of diffusion of the growth units (solute molecules) to the surface of the existing nuclei and their incorporation into the structure of the crystal lattice. Later (Section 4.5) we will show how the habit of a crystal is related to the internal arrangement of molecules.

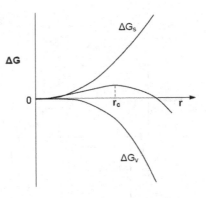

Variation of ΔG with the size of the critical nucleus.

The change in variation of Gibbs free energy of a *nucleus* versus its radius, *r* (assuming a spherical shape for the nucleus) can be given by the following equation.

$$\Delta G(r) = -\underbrace{\left(\frac{4\,\pi r^3}{3\,\Omega}\right) k_B T \ln(1+\sigma)}_{\Delta G_V} + \underbrace{4\,\pi r^2 \gamma}_{\Delta G_S}$$

r = size of the cluster, Ω = volume occupied by cluster, σ = relative supersaturation, k_B = Boltzmann constant and γ = solid-liquid interfacial energy.

$$r_c = \frac{2\,\Omega\,\gamma}{k_B \ln(1+\sigma)}$$

The critical nucleus, r_c, corresponds to the maximum in the curve. It signifies the minimum amount of a new phase capable of independent existence.

$$\Delta G^* = \frac{16\,\pi\,\Omega^2\,\gamma^3}{3\,[k_B\,T\,\ln(1+\sigma)]^2}$$

The maximum free energy of formation, ΔG corresponds to the activation energy barrier for nucleus formation.

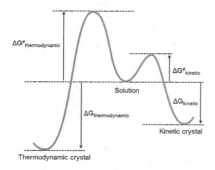

4.3   Thermodynamics and Kinetics of Crystallization

In the perspective of crystal engineering, it is convenient to describe the thermodynamics and kinetics of crystallization in terms of intermolecular interactions. Stabilization of a crystal with a unique packing can be described through a simple energy profile diagram. The high energy intermediate (nucleus) can be considered as a supramolecular transition state involving aggregation of a critical number of molecules or molecular aggregates. It is difficult to address how a system chooses a specific pathway in the supramolecular reaction from aggregating molecules to a nucleus that eventually stabilizes a crystal with a particular packing. Retrosynthesis (Section 3.4) is a useful tool to hypothesise a liquid-like structure for this state; it acts as a bridge between aggregation in the solution and the solid state structure observable by different techniques. However, we can consider the crystal packing of molecules for example, from two different viepoints: a geometrical approach originally proposed by Kitaigorodskii that considers weak interactions between molecules but de-emphasises directionality. This model emphasises close packing or space minimization to achieve the lowest energy. The other, more chemical, approach is based on directional interactions conceivable through supramolecular synthons (Chapters 2 and 3). In the former, the system optimises close packing by sampling all possible modes of aggregation between interacting molecules before selecting the one lowest in energy, possibly the global minimum. The result can be considered as a thermodynamic crystal or polymorph (Chapter 5). In the chemical approach, the individual interactions are very important. Once a strong synthon forms, it cannot be undone, and the system proceeds onwards with the formation of new and weaker synthons. This is the kinetic crystal or at least one of the kinetic outcomes of crystallization. Because of this kinetic versus thermodynamic dichotomy, we may encounter the phenomenon of polymorphism.

Two possibilities need to be considered. In the first, thermodynamic and kinetic outcomes of crystallization are identical. For example, crystals of benzoic acid, urea, naphthalene, and D-glucose are monomorphic because only one set of interactions (represented by synthons, be they dimer, stacking or hydrogen bonding) dominate the aggregation in solution. In the second case, the kinetic forms are different from the thermodynamic crystal, and polymorphs may be observed with a greater or lesser degree of ease, provided adequate experimentation is carried out. If a solution is supersaturated, it will eventually crystallize. The critical size controls the probability of the nucleus formation on any given timescale.

A major approximation involved in classical theory of nucleation is that the arrangement of molecules in the crystal nucleus is identical to that in solute crystal. In other words, the surface free energy of the nuclei will be equal to that on the crystal interface. This is unlikely in many instances as the energy barrier to a less stable state (local minimum in the energy profile diagram) is lower than when going to the most stable state (global minimium). This is the basis of *Ostwald's Rule of Stages* which suggests that the pathway to the final crystalline state will pass through all less stable states in order of increasing stability. Whether or not this rule holds always is difficult to predict. A major hurdle in studying nucleation and the early stages of growth of the nucleus is that currently available experimental methods can hardly access its size (1–100 nm range) and its periodic structure which falls in the range of a few thousands of molecules.

Free energy minimization is not a necessary criterion for crystal morphology. The simple model proposed by Gibbs provides a minimum free energy barrier for nucleation and hence a nucleation pathway of fastest kinetics. The Gibbs model oversimplifies the dependence of the structure of the crystal nucleus on a single variable, r. In reality, we need to consider a multidimensional surface to account for the free energy landscape. This suggests that multiple pathways exist between the initial state where individual molecules are suspended in a solvent, and the final state with crystals greater than the critical size driven by ΔG. In this simple model, the observed rate of nucleation depends solely on the fastest nucleation pathway.

4.4   Crystal Growth

Once a crystal is stabilized from the *critical nucleus*, the growth units (building blocks) can diffuse from the surrounding supersaturated solution to the surface of the critical nucleus and incorporate themselves onto

In situ cryocrystallization

There are many organic compounds which are liquids at or around room temperature. These include solvents, low melting solids and ionic liquids. These substances solidify at low temperatures and will often yield single crystals, but it is normally difficult to handle such crystals for X-ray study. Recent advances in *cryocrystallography* have made it possible to grow single crystals in situ at low temperatures and collect X-ray data. The crystal structures of these substances provide a valuable addition to the information bank of the crystal engineer. In a typical technique, crystallization is conducted by blowing a cold stream of N_2 onto the substance in a quartz capillary mounted on the diffractometer. Some amount of microheating/ annealing is often required to produce a single crystal from the microcrystalline mass that is generally obtained. Almost all the commercial diffractometers are equipped with a low temperature device which makes cooling of neat liquid samples or saturated solutions feasible. This technique was pioneered by Roland Boese.

Aristotle's famous philosophy says that "No reaction occurs in the absence of a solvent". This philosophy had a big influence in the development of synthetic chemistry and many reactions were performed in the presence of solvents even if there was no reason to do the reaction in solvent media.

Crystal energy landscape

Many techniques are used to crystallize organic molecules. One encounters many solid forms like polymorphs, solvates and multi-component solids. Computational methods are being developed to predict the thermodynamic feasibility of such crystallization processes. In this context, the *crystal energy landscape* or *structural landscape* provides a framework to compare the most easily accessed structures in a given system. Often, there will be minima that are compatible only with certain types of intermolecular interactions. The crystal energy landscape is a useful tool in the understanding of the organic solid state.

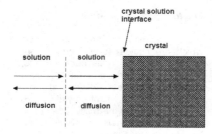

the crystal resulting in growth. How fast the nuclei grow depends on the rates at which growth units attach and/or remove themselves from the nuclear surface. The rate-controlling step for growth can be the diffusion of species through the reaction zone and the solution, in other words the rate of the surface reaction. Given the slow rate of diffusion of species through the reaction zone, the thickness of this zone will at some point become sufficiently large so that diffusion through the reaction zone becomes the rate-determining step. By opting for selected surface active species, it is possible to control the dimensions of growth rate. Formation of nanocrystals and biomineralization are the result of controlling crystallization at the reaction zone. The presence of surface active groups or proteins, may provide a barrier for limiting the dimension of the growing nucleus — a compromise between building a bulk crystal and aggregation around the surface of the crystal.

4.4.1   *The Terrace-Ledge-Kink Model of Crystal Growth*

The terrace-ledge-kink model proposed by Walther Kossel provides a simplified picture (also referred as the Kossel-Stranski model) for understanding crystal growth. As the nucleus grows, the high energy sites on its surface are minimized. In other words, stable growth occurs at sites on the crystal nucleus where newly grown structures can maximize their contact with the crystal surface. On the molecular scale, a crystal surface contains many kinds of interfaces between the growing crystal nucleus and the crystallizing solution. The adjacent diagram illustrates the mechanism of crystal growth according to the Kossel-Stranski model. The growth unit is a small cubic block. The flat region of a surface is called a *terrace* and the raised partial layer is referred as a *step*. The growth unit makes minimal contact with the crystal surface. The growth at this position is not so stable and the dissolution of the block by the solvent is more likely. The block may attach to a *ledge* site where two surfaces are available for interaction providing better stability. The growth is the most favourable at a *kink site* where the growth unit gets maximum contact (highest coordination) and hence greatest stability — and least probability of dissolution. Also, incorporation of a block at a kink site will lead to a new one, thus producing a repeatable step in crystal formation. For a given solute concentration, the rate at which molecules are attached to the crystal surface will depend on the kink density. It is possible to alter the growth rate by selectively blocking kink sites (Section 4.6). The strength of the interactions and their anisotropy will determine the equilibrium kink density. If the interactions in a crystal are stronger parallel to the step edge than perpendicular to it, the step will have a low kink density. As a consequence the rates will vary in both directions.

4.4.2 *Two-dimensional Nucleation versus Growth at Dislocations*

The terrace-ledge-kink model assumes the pre-existence of steps on a crystal surface. Growth will not take place unless new steps are created. In such a situation, two possibilities arise. In the first case, a two-dimensional crystal nucleus must exist on a surface. In the second case, dislocations provide the source for new steps and kink sites. The step-growth will continue until the whole surface is completed. For further growth to occur, a new two-dimensional nucleus needs to be generated. Higher supersaturation levels are favourable for two-dimensional growth since it is difficult to generate a new nucleus on an already existing flat crystal surface. At low supersaturations, however, the critical size is large and the probability of creating 2D nuclei decreases. Fortunately crystals are not perfect and they contain dislocations which act as steps. Screw dislocations are a continuous source of new steps and provide a way for steps to grow uninterrupted.

Crystal growth increases the sizes of existing nuclei. Thermodynamics favors the formation of larger nuclei at the cost of smaller nuclei which redissolve. Overlap of nucleation and growth rates decide the nature of the crystals formed. Pure raw materials are often produced to required sizes through precipitation from aqueous solution. Large single crystals for specialty applications are grown either from pure molten solids or by solution evaporation. Rapidly cooled liquids often skip crystallization and end up as amorphous solids. Generally, the size of a precipitated particle increases if the reaction is allowed to occur in the presence of previously precipitated particles. The continuous conversion of small particles into larger ones is mediated by agglomeration under conditions close to equilibrium.

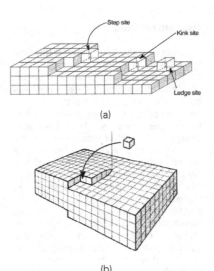

Illustration of growth process on (a) two-dimensional surface and (b) screw dislocation.

Ostwald ripening

Ostwald ripening has a rational physical basis. A less stable phase is more soluble, that is σ is smaller. Nucleation of a less stable phase means γ^3/σ^2 will be smaller. This suggests that γ must be reduced by an even greater factor. Remember that γ, the interfacial energy, is the difference between the change in free energy in forming a crystal with a surface and in an infinite crystal. The free energy terms ΔG_V and ΔG_S in turn depend on changes in enthalpy and entropy of the bulk and surface respectively. The phases corresponding to local minima are probably entropy driven. Since the surface term is more susceptible to change than the bulk term during nucleation, the total free energy change will depend strongly on γ, which in turn depends on the surface entropy.

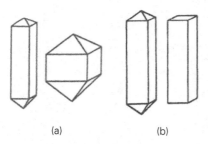

(a)   (b)

(a) Same morphology but different habits. The relative dimensions of the crystals vary.

(b) Different morphologies but similar habits. The crystal faces are different but the relative dimensions are similar.

4.4.3   *Ostwald Ripening*

Closely related to this is a phenomenon called *Ostwald ripening*, in which large crystals grow at the expense of smaller crystals. This spontaneous process occurs close to equilibrium and takes place because larger crystals are more energetically favored than smaller ones. It is easier to nucleate many small crystals (kinetically favored). However, small crystals have a greater surface area to volume ratio. Also, the molecules on the surface are energetically less stable than the ones already well ordered and packed in the interior. In comparison, large crystals with their greater volume to surface area ratio, represent a lower energy state. Hence, large crystals can be considered to be thermodynamically favored. Ostwald ripening is the tendency of small crystals to transform into a few large ones. You may wonder why this phenomenon does not happen more often. A plausible explanation is that the simultaneous nucleation of several small crystals reduces the level of supersaturation thereby disfavoring the growth of large crystals.

4.5   Crystal Morphology and Habit

The terms *morphology* and *habit* refer to descriptions of the external form of a crystal. Crystal morphology refers to the faces of a crystal and the interfacial angles. Morphology is arguably the most obvious characteristic of a crystal, be it for selection for crystal structure determination or for the reliable identification of polymorphs. The crystal habit is a more qualitative descriptor and describes the shape, say prism, needle or plate. It is based on the relative lengths of the major dimensions of the crystal. The accompanying figure shows how two crystals can have the same morphologies but different habits. Morphology and habit are important for processing in industries. Filtering, drying and free flowing ability depend on the shape and size distribution of the crystalline substance.

Crystal growth involves the transportation of growth units to the surface of the crystal. Two factors affect growth: the rate at which units migrate to the surface via diffusion and the rate of reaction at the surface. The crystal faces that develop are related to the internal structure. If crystals grow nearly uniformly in all three dimensions, they become isometric. For example, an octahedral shape is associated with internal cubic symmetry. When growth dominates on a single plane, the habit is flaky or platy. Crystal growth in one direction will result in a columnar or needle-like form. As a general rule, the slowest growing faces dictate the final morphology and habit.

4.5.1 *Crystal Morphology and Crystal Symmetry*

The symmetry observed in crystals as exhibited by their faces is due to the ordered internal arrangement of atoms. A lattice is defined by the spacings between points and the directions (or angles) between the points. A crystal face develops along planes defined by the lattice points. In other words, all faces must intersect molecules that make up the points. Four important features that relate a crystal lattice to its morphology: (i) A face is more commonly developed in a crystal if it intersects a large number of lattice points. This is known as the *Bravais Law*; (ii) The angles between various crystal faces is determined by the spacing between lattice points; (iii) Since all crystals of the same solid have the same spacings between lattice points, the angles between corresponding faces in all crystals of that substance will be the same. This *Law of Constancy of Interfacial Angles* was discovered by Nicholas Steno in 1669, and was a key development in the history of crystallography; (iv) The symmetry of the lattice determines the angular relationships between crystal faces.

4.6 Crystal Morphology Engineering

Besides the external parameters, temperature and supersaturation, the presence of impurities in the crystallization medium affects crystal growth. Crystal morphology and habit are important characteristics of powdered materials. They affect the ease with which a solid can be pressed into a tablet (Section 6.6.2.2). Equidimensional crystals (equant) are usually preferred in the pharmaceutical industry as they have better handling and processing characteristics such as flowability and compactability.

Crystal morphology engineering is therefore a valuable tool to tailor technological properties of solid materials. Control of crystal morphology can be achieved either by solvent selection and/or tailor-made additives. Tailor-made additives are employed for the

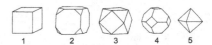

The illustrations show how a perfect cube 1 is progressively changed into the octahedron 5. The cube 1 be changed into 5 stepwise by slicing planes parallel to (111) on all eight corners. Notice the change in crystal faces; the corresponding Miller indices will also vary. These five shapes possess the same set of symmetry elements as the simple cube and are thus various crystal habits.

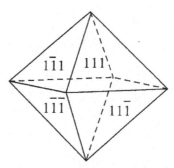

The figure shows Miller-Bravais indices for faces {111} in the cubic system. The faces are all equivalent and the overall shape will be cubic or octahedral. Crystal faces represent limiting surfaces of growth. A habit refers to a set of symmetrically equivalent faces. The symbol { } indicates a form. The multiplicities of the form {111} in the monoclinic, orthorhombic and tetragonal systems are all different and also distinct from that in the cubic system.

control of nucleation and growth of molecular crystals. *Tailor-made inhibitors* of crystal growth can be used for a variety of purposes, which include assignment of absolute structure of chiral molecules and polar crystals, and crystallization of a desired polymorph (for example ibuprofen and paracetamol).

In the context of pharmaceutical solids, solvent-induced habit modification is limited by solvent toxicity and cost, crystallization efficiency and purity requirements of the final product. Hence, the use of additives as crystal habit modifiers of APIs is usually preferred. Environmental concerns can generally be overcome by employing pharmaceutically accepted excipients as crystal habit modifiers.

The habit of a crystal is defined by the relative rates of growth of the crystal in different directions. The faster the growth in a given direction, the smaller the face developed perpendicular to that direction and vice versa. There is a correlation between the unit cell parameters and the rates of crystal growth in particular directions. Crystal growth is faster in those directions that are associated with shorter unit cell axes. Crystals in which one of the unit cell lengths is much smaller than the other two tend to grow as needles. Crystals in which one of the unit cell lengths is much larger than the other two tend to grow as plates. Crystals in which the three unit cell lengths are nearly the same (say cubic crystals) tend to grow equidimensionally. The most popular model for morphology prediction is called the Bravais-Friedel-Donnay-Harker (BFDH) model, and the CSD can be used to calculate the expected BFDH morphology of a crystal from the crystallographic information that is present in the database. The dramatic morphological changes associated with the growth of organic crystals in the presence of additives reveal the high degree of specificity of the interactions formed by the added molecule with the different structured surfaces of the crystalline matrix. Therefore, morphological changes have a direct bearing on the mechanisms of the adsorption–inhibition process at the molecular level.

4.6.1   *Tailor-made Inhibitors*

It is possible to use the concept of molecular recognition at interfaces to address open questions in the areas of crystal nucleation and growth, crystal polymorphism, and interactions with the growth environment.

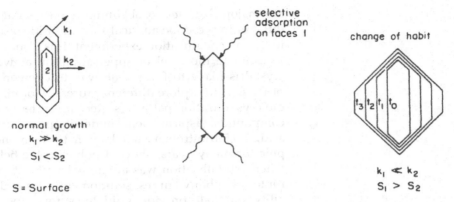

A schematic representation of how relative growth rates change crystal habit.

Crystallization of benzamide, $C_6H_5CONH_2$, in the presence of benzoic acid as an additive is an example where the host-additive interactions at the growing crystal-solution interface can be highlighted and correlated with changes in crystal morphology. Benzamide crystallizes from ethanol as plates in the space group $P2_1/c$. In the crystal structure, the homosynthons are further interlinked by hydrogen bonds to yield ribbons. The ribbons are stacked along the b-axis to form stable (001) layers. These layers pack along the c-direction *via* weak $\pi\cdots\pi$ interactions between phenyl groups, thus accounting for the {001} plate-like shape of the crystals. The additive, C_6H_5COOH, can easily replace a molecule of benzamide at the end of a ribbon by forming a strong heterosynthon. This inhibits the growth of the benzamide crystals along b, thus transforming the pure plate-like crystals into needles elongated along the a-axis.

The additive benzoic acid inhibits the growth of benzamide along the b-axis.

4.7   Why is it that all Compounds don't seem to Crystallize Equally Well or Equally Quickly?

Crystallization is commonly employed for purification of organic compounds. There are a few solids such as benzoic acid which will separate as pure phases even when the solution is largely impure. It is well-known that organic compounds in particular show considerable variation in its crystallization behavior. Conformational flexibility and competitive interactions are major obstacles to obtaining good crystals. It is difficult to prescribe an ideal solvent or ideal solubility for a crystallization experiment. It is also difficult to estimate the level of supersaturation at which a crystal is likely to form. One way to overcome this problem is to explore different processes for mapping the crystallization pathways. Recently, a few organic compounds (aspirin, benzamide, ibuprofen, maleic acid, 1,3,5-trinitrobenzene) known to be monomorphic for many years, showed polymorphic behavior when crystallization was attempted with alternative routes. Solubility curves, solution eutectics and solubility data all provide guidelines for choosing a particular method. However, poor crystallization may arise due to many reasons. Problems arise due to failure to nucleate, uncontrolled nucleation, solvation or hydration, and strong affinity for impurities. The student is advised to distinguish between poor crystallization and slow crystallization. A compound may crystallize very slowly but it may form good crystals in the end. This could happen because the molecules are conformationally flexible. Other compounds may crystallize quickly but the crystals formed may be bad. This could arise because nucleation is not controlled. There are still others that form good crystals quickly under a variety of conditions. The study of crystal growth is important from both fundamental and practical viewpoints. Much has been done with ionic and inorganic solids. Crystal growth of molecular solids is a new subject that has considerable overlap with crystal engineering.

4.8 Summary

- Crystallization is fundamental to many natural, chemical and industrial processes.
- Understanding nucleation and crystal growth are inherent to the process of crystallization.
- Metastability or supersaturation/supercooling/super condensation is a prerequisite for initiating crystallization from solution, melt, vapor or a solid.
- Selected crystallization techniques for molecular solids are solvent evaporation, melt cooling, sublimation, single crystal to single crystal transformations and mechanochemistry (grinding).
- An important concept in nucleation is the size of the critical nucleus and how it depends on interfacial energy (crystal structure variable) and supersaturation levels (solution variable).
- A simple energy profile diagram (reaction coordinate) provides pathways for the passage of interacting molecules in the crystallization medium towards the nucleus (the supramolecular analog of the transition state) and eventually to a thermodynamic or

kinetic crystal. When thermodynamic and kinetic outcomes of crystallization are identical, polymorphism is not easy to observe. When they are not, polymorphs can, in principle, be isolated.

- Kinetic crystals are formed because of the establishment of directional intermolecular interactions and the supramolecular synthons that are conserved wherefrom.
- Thermodynamic crystals are obtained by a minimization of the total free energy of the system. If necessary, directionality of intermolecular interactions may be compromised in such crystals.
- Ostwald's rule of stages suggests that the pathway to the final crystalline state will pass through all metastable states in order of increasing stability.
- The Kossel-Stranski terrace-ledge-kink model provides a simple description of the growth of a crystal on a surface. Two dimensional crystal nuclei and dislocations provide surfaces for growth, depending on the concentration of supersaturation.
- Ostwald ripening is the process in which small crystals dissolve and reappear as larger crystals.
- The morphology of a crystal is a quantitative descriptor of the crystal faces and the angles between these faces.
- The habit of a crystal is a qualitative shape descriptor.
- Impurities can affect the morphology of crystals. Crystal engineering concepts can be exploited to alter the growth of a solid for achieving a favorable physical property by the addition of deliberate additives, *tailor made inhibitors*.

4.9 Further Reading

Books

A. S. Myerson, D. A. Green and P. Meenan (eds), *Crystal Growth of Organic Materials*, 1996.

H.-H. Tung, E. L. Paul, M. Midler and J. A. McCauley, *Crystallization of Organic Compounds. An Industrial Perspective*, 2009.

Papers

A. Collet, M. J. Brienne and J. Jacques, Optical resolution by direct crystallization of enantiomer mixtures, *Chem. Rev.*, 80, 215–230, 1980.

I. Weissbuch, L. Addadi, M. Lahav and L. Leiserowitz, Molecular recognition at crystal interfaces, *Science*, 253, 637–645, 1991.

L. Addadi and S. Weiner, Control and design principles in biological mineralization, *Angew. Chem. Int. Ed.*, 31, 153–169, 1992.

S. Mann, D. D. Archibald, J. M. Didymus, T. Douglas, B. R. Heywood, F. C. Meldrum and N. J. Reeves, Crystallization at inorganic-organic interfaces — Biominerals and biomimetic synthesis, *Science*, 261, 1286–1292, 1993.

J. Hulliger, Chemistry and crystal growth, *Angew. Chem. Int. Ed.*, 33, 143–162, 1994.

F. Schüth, Nucleation and crystallization of solids from solution, *Curr. Opin. Sol. State Mat. Sc.*, 5, 389–395, 2001.

B. A. Garetz, J. Matic and A. S. Myerson, Polarization switching of crystal structure in the nonphotochemical light-induced nucleation of supersaturated aqueous glycine solutions, *Phys. Rev. Lett.*, 89, 175501, 2002.

R. Buller, M. L. Peterson, Ö. Almarsson and L. Leiserowitz, Quinoline binding site on malaria pigment crystal: A rational pathway for antimalaria drug design, *Cryst. Growth Des.*, 2, 553–562, 2002.

M. Sakamoto, Spontaneous chiral crystallization of achiral materials and absolute asymmetric transformation in the chiral crystalline environment, in *Enantiomer Separation* (ed. F. Toda), 2004.

S. Parveen, R. J. Davey, G. Dent and R. G. Pritchard, Linking solution chemistry to crystal nucleation: The case of tetrolic acid, *Chem. Commun.*, 1531–1533, 2005.

A. V. Trask and W. Jones, Crystal engineering of organic cocrystals by the solid-state grinding approach, *Top. Curr. Chem.*, 254, 41–70, 2005.

T. Friščić and L. R. MacGillivray, Single-crystal-to-single-crystal [2+2] photodimerizations: from discovery to design, *Z. Kristallogr.*, 220, 351–363, 2005.

F. H. Herbstein, On the mechanism of some first-order enantiotropic solid-state phase transitions: from Simon through Ubbelohde to Mnyukh, *Acta Cryst.*, B62, 341–383, 2006.

A. Ramanan and M. S. Whittingham, How molecules turn into solids: the case of self-assembled metal-organic frameworks, *Cryst. Growth Des.*, 6, 2419, 2006.

D. Erdemir, A. Y. Lee and A. S. Myerson, Nucleation of crystals from solution: Classical and two-step models, *Acc. Chem. Res.*, 42, 621–629, 2009.

M. B. Hursthouse, L. S. Huth and T. L. Threlfall, Why do organic compounds crystallise well or badly or ever so slowly? Why is crystallisation nevertheless such a good purification technique? *Org. Proc. Res. Dev.*, 13, 1231–1240, 2009.

W. L. Noorduin, E. Vlieg, R. M. Kellogg and B. Kaptein, From Ostwald ripening to single chirality, *Angew. Chem. Int. Ed.*, 48, 9600–9608, 2009.

S. Mahapatra, K. N. Venugopala and T. N. Guru Row, A device to crystallize organic solids: Structure of ciprofloxacin, midazolam, and ofloxacin as targets, *Cryst. Growth Des.*, 10, 1866–1870, 2010.

I. Halasz, Single-crystal-to-single-crystal reactivity. Gray, rather than black or white, *Cryst. Growth Des.*, 10, 2817–2823, 2010.

Web links

Paul D. Boyle, North Carolina State University, http://www.xray.ncsu.edu/GrowXtal.html
Maarten Dinger and Jerzy Klosin, University of South Florida, http://xray.chem.ufl.edu/growing tips.htm

4.10   Problems

1. The space group of crystals of D-glucose is $P2_12_12_1$. Write down the space group of crystals of L-glucose.

2. The compound [Ni(4,4'-dipy)(ArCOO)$_2$(MeOH)$_2$] is a coordination polymer that crystallizes in the space group $P4_12_12$ (or $P4_32_12$). Show that this space group preference leads to a particular ligand stereochemistry for this polymer.

3. Crystal data for three crystals are as follows: (i) Monoclinic, $C2/c$, $a = 26.584$ Å, $b = 6.062$ Å, $c = 11.318$ Å, $\beta = 108.038°$; (ii) Cubic, $P2_13$, $a = 17.869$ Å; (iii) Triclinic, $P\bar{1}$, $a = 3.804$ Å, $b = 10.502$ Å, $c = 11.112$ Å , $\alpha = 77.84$, $\beta = 84.26$ $\gamma = 80.17°$. These crystals are equant, platy or needle-like. Say which is which.

4. The kinetic polymorph of 1,3,5-trinitrobenzene is considerably less stable than its thermodynamic polymorph. Yet, it is the form that is routinely obtained and it remains stable indefinitely at ambient conditions. Under what conditions are kinetic polymorphs stable?

5. Curcumin is the principal ingredient in the popular Indian spice turmeric and is responsible for its characteristic yellow to bright orange color. Crystals of a new

polymorph were obtained upon attempted co-crystallization of curcumin with 4-hydroxypyridine. Such observations have been made frequently in the past for several types of compounds but there is no proper record. Suggest reasons why the addition of another compound to a crystallization mixture can result in the formation of a new polymorph.

Polymorphism

5

Polymorphism is the phenomenon in which the same chemical substance exhibits different internal crystal packing arrangements. The word *polymorph* is derived from the two Greek words *poly*, meaning many and *morph* meaning form. The word *morph* sounds similar to the word *morphology*, which refers to the shape or habit of a crystal. Therefore we need to carefully distinguish between different habits or crystal shapes that enclose what is essentially the same crystal structure, and *polymorphism* which is the occurrence of crystals that have completely different internal structures. To expand this idea further, the same compound adopting the same crystal structure can crystallize in different shapes or habits, such as needles and plates. These forms do not constitute polymorphs but only different morphologies. Two polymorphs, on the other hand, have entirely different crystal structures. The mutual arrangements of molecules, atoms or ions are different. The physical and chemical properties of polymorphs can also be substantially different. Different polymorphs of a compound are often obtained under different crystallization conditions. Different morphologies or crystal habits are also often obtained under different crystallization conditions. To add to the confusion, different polymorphs often have different morphologies. All this makes it difficult to distinguish polymorphism from simple habit differences.

5.1 What is Polymorphism?

Polymorphs are crystals in which a chemical compound takes different arrangements of molecules. *Polymorphism is therefore an exclusively solid state phenomenon.* Owing to differences in crystal structure, polymorphs can often have different morphologies, solubilities, color, melting or sublimation temperatures, densities, thermal or electrical conductivities and other physical properties. Polymorphism is observed in all crystalline compounds: molecular and non-molecular, organic, inorganic, organometallic and coordination complexes. Eilhardt Mitscherlich, in 1820, was the first to recognize polymorphism when he identified different crystal structures for sodium arsenate phosphate. In 1832, Friedrich Wöhler and Justus Liebig discovered the first example of polymorphism in an organic solid, benzamide. With the development of analytical techniques, the characterization of polymorphic structures and their crystallization behavior have become easy and a more detailed study of the phenomenon of polymorphism has become possible. Polymorphs have different crystal structures. Therefore all types of diffraction techniques on single crystals and powders are of great importance to the study of the phenomenon. In 1938, J. M. Robertson (see Section 1.1) and Alfred Ubbelohde used X-ray crystallography and determined the crystal structure of the dimorphs of resorcinol, or 1,3-dihydroxybenzene. This is possibly the first example of a molecular solid whose polymorphs were characterized by full crystal structure determinations.

5.1.1 *Polymorphism and the Pharmaceutical Industry*

Although differences in crystal morphology and differences in crystal structure (polymorphism) are distinct phenomena, there are still many connections. Since polymorphs have different crystal structures, it is likely that they also have different morphologies. Indeed, in its initial days, two centuries or so ago, polymorphism was mostly deduced from crystal morphology. Early examples were generally restricted to well-developed crystals, mainly minerals and other inorganic substances. Today, polymorphism of organic substances is widely investigated. Some typical examples of polymorphic solids that find applications in daily life are indigo, cocoa butter, sorbitol, lead tetraethyl and copper phthalocyanine. In the context of the pharmaceutical industry, differences in crystal morphology (Section 4.5) can affect processing properties like filtering, drying, flow, tableting, rate of dissolution, shelf life and bioavailability. So, the process of crystallization and its effect on the crystal morphology are of much importance to the drug manufacturing industry.

A very significant number of marketed drugs exhibit polymorphism under experimentally accessible conditions. There are two main reasons for this. The first reason is structural: most drug molecules contain a number of functional groups that are both capable of forming hydrogen bonds and are flexible. This combination of a functional group, containing rotatable bonds, which can also donate or accept hydrogen bonds can lead to several possible orientations of molecules in the crystal, giving different crystal structures. Such substituent groups include the commonly found OH, NH_2, $NHCOCH_3$, $COOCH_3$ and $NHCH_3$. The second reason is thermodynamic: the demand for maximum yield and high production rates forces chemical and pharmaceutical industries to operate crystallization processes far from equilibrium. Under these conditions, there is a marked tendency to form polymorphs.

Independent of this, there is a compelling commercial reason for the production of pure polymorphic phases in the pharmaceutical industry. In the early 1990s, it was held in the US law courts that a new polymorph of a drug may, under certain circumstances, be held as a distinct legal entity and as such, entitled to separate patent protection. A mixture of polymorphs is chemically pure but it is not pure in a crystallographic sense because it consists of crystals with different crystal structures. The legal ruling, in effect, recognized that crystallographic purity is a separate issue from chemical purity and that a drug has both chemical and crystallographic properties that may be independently entitled to patent protection. This ruling, which was made specifically with respect to the dimorphs of the block buster anti-ulcer drug ranitidine (Section 5.7.1) had far reaching legal and scientific implications. For a start, there is now a high motivation among pharmaceutical manufacturers to find optimum conditions for the production of one or another polymorph of a drug.

Two common scenarios quickly emerged. An innovator company discovers a drug and patents it as a particular polymorph. A second company (called a generic) finds a second polymorph and patents it as a new crystal form. If the generic can make the second polymorph, completely uncontaminated by the first one, it can sell it. In the second scenario, the

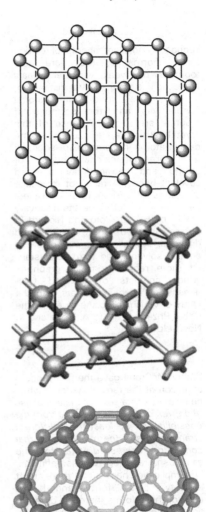

Like molecules, elements such as B, C, Sn, P, As, S, Se can also crystallize in different ways. Polymorphs of crystalline elements are usually referred to as allotropes rather than as polymorphs. Till recently, only two allotropes of carbon (graphite and diamond) were known. Graphite consists of planar hexagonal layers consisting of an infinite number of fused six-membered rings with d(C–C) = 1.421 Å stacked in an ABAB

(Continued)

(Continued)

sequence. Adjacent sheets of carbon atoms normal to (001) are held together by weak van der Waals interactions separated by a distance of 3.34Å. The two dimensional conductivity and the softness of graphite are a consequence of this crystal structure. In contrast, the carbon atoms in diamond are arranged in a three-dimensional network of strong and tetrahedrally directed covalent bonds. A range of molecular allotropes of carbon (the fullerenes) have been isolated since the discovery of C_{60} by Harold Kroto and Richard E. Smalley in 1985. Unlike graphite and diamond, the fullerenes are soluble in organic solvents and form co-crystals. Of more recent origin is graphene which consists of isolated two-dimensional graphitic sheets, and the study of which led to the award of the Nobel Prize in 2010 to A. Geim and K. Novoselov.

Walter McCrone gave the first modern definition of the term *polymorph* when he termed it as *a solid crystalline phase of a given compound resulting from the possibility of at least two different arrangements of the molecules of that compound in the solid state.* Note the cautious phraseology of this definition. The implication here is that there are at least two crystal forms, and that in each of these forms there is a different arrangement of molecules of the compound.

Polymorphs are numbered traditionally in the order in which they have been discovered. Accordingly, different crystal forms of a compound may be called A, B, C or α, β, γ or 1, 2, 3.

innovator company discovers a second polymorph of its drug and patents it, even as its patent on the first form is expiring. After the patent on the first form expires, generics may make and sell it, but what they sell should be free of the second form, for which a valid patent is held by the innovator. Both innovators and generics therefore have strong motivations to find new polymorphs of a drug. The innovator would like to find all possible polymorphs and patent them all. The generic would like to carve out some intellectual property in the form of a new polymorph. Alternatively, the generic would like to make the first polymorph completely uncontaminated by protected forms. There is therefore an intense desire on the part of both innovators and generics to be able to detect very small quantities of one polymorph in a solid material that consists predominantly of another polymorph.

Crystal engineering provides an ideal platform to combine academic curiosity and industrial challenges to explore polymorphism. In this chapter we confine our discussion to polymorphs. Chapter 6 addresses issues related to solvates, hydrates, co-crystals and salts. All these variations together with polymorphs constitute the structural landscape that is associated with any particular organic molecular solid.

5.1.2   *Some Simple Definitions*

Polymorphism is the phenomenon wherein the same chemical compound exhibits different crystal structures. These different structures or forms are called polymorphs. Let us look more closely at the words *same* and *different* in the above definition. Normally, there is no confusion in distinguishing between two compounds and in saying whether or not they are the same. Again, there is usually no ambiguity as to whether or not two crystal structures are different. However, there are some situations when the interpretations of the words *same* and *different* lead to ambiguities. In these cases, it is sometimes problematic to determine whether or not polymorphism is present.

In *conformational polymorphism*, each of the isolated crystal forms contains a different conformation of the same molecule. The energies of the non-bonded interactions that govern intramolecular conformation are comparable to the energies between molecules. There

is interplay therefore between intramolecular conformations and intermolecular orientations. It is difficult
to say which controls which; in some systems, there
are a large number of conformational polymorphs that
reflect the changing balance between these two influ-
ences. Different conformers, of course, must have
different crystal structures. These conformations are
considered to arise from the same chemical compound
if they are in facile equilibrium at the temperature of
the experiment. If the conformations, however, are
locked in (frozen) at the temperature of the experiment, then they would be taken to be different
chemical compounds, and their crystal structures
would not be polymorphs.

This is a very general principle. Two variations of
the same substance that do not interconvert in solution
or in the melt are not considered to be the same chemical compound. As such, their crystal structures are not
considered to be polymorphs. Accordingly, crystal
structures of the R and S enantiomers of a compound,
which owes its chirality to the presence of an asymmetric carbon atom, are not polymorphs. Similarly,
crystal structures of resolved (R or S) and racemic (R, S)
forms are not polymorphs. For molecules that are chiral
because of restricted rotation (biphenyls), the chiral
forms are polymorphic if the conformers interconvert
in solution: if they do not, their crystal structures
are not polymorphs. By the same token, crystal forms
of tautomers are usually considered to be polymorphs, because the tautomers generally equilibrate
in solution at the temperature at which the solid forms
are isolated. This phenomenon is called *tautomeric
polymorphism*. An interesting example is shown by the
drug omeprazole and this is explained in Section 5.7.4.

The concept of molecular sameness needs to be carefully considered in coordination polymers, which are
compounds in which organic ligands act as connectors
between inorganic modules (Chapter 7). A pair of
coordination compounds can have the same chemical
formula but different structures. The differences can
arise from differences in geometry of the metal ions,
the ligands, or because of other factors like ionization and solvation. In these compounds, concepts
such as "same chemical substance" or "same molecule" are not always clear. Accordingly, it is
sometimes difficult to say if two different crystal
structures are polymorphs or whether they are
crystal structures of different compounds. The term

Some compounds that crystallize as
conformational polymorphs.

Tautomeric interconversion in omeprazole.

Optical micrographs of 3-acetyl-
coumarin: (i) Left. Yellow prisms (Form
B) from glacial acetic acid; (ii) Center.
Concomitant mixture of forms A and B
obtained at 278 K; (iii) Right. White
needles (Form A) from 1:1 CHCl$_3$:*n*-
hexane obtained at 298 K.

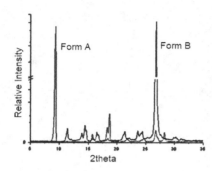

PXRD patterns of concomitant polymorphs A and B of 3-acetylcoumarin.

Disappearing polymorphs

The steroid progesterone is known to be polymorphic for 70 years. Two polymorphs were characterized crystallographically. Form I is the stable form. Form II is metastable and converts rapidly to Form I. However, a 50 year old sample of Form II held in the University of Innsbruck, still continues to be stable. A closer examination of these crystals has shown that they contain unidentified impurities. Our inability to produce long-lived samples of Form II today may thus be on account of changes in the impurity profile. This example casts a different light on the phenomenon of disappearing polymorphs, which in the end is a scientific contradiction.

supramolecular isomerism has been used, and this may not always be the same as polymorphism. This matter is discussed in more detail in Chapter 7.

Two crystal structures are said to be different when they give different diffraction patterns. However, because of advances in modern instrumentation, it is now possible to observe small differences in diffraction patterns of some crystals. Do these crystal forms qualify as polymorphs? Such crystals may be small modulations of one another and there are many borderline situations and grey areas. A study of these examples is in the research forefront today. Some interesting examples are given in Section 5.7.

By concomitancy is meant the phenomenon in which different polymorphs are obtained in the same experiment. *Concomitant polymorphs* are very similar in energy and there needs to be a fine balancing of various thermodynamic factors and solvent–solute interactions in order for concomitancy to be even observed. In dramatic cases, the concomitant polymorphs have different colors or morphologies and the phenomenon is easily detected.

Disappearing polymorphs are crystal forms that fail to reappear after their initial isolation. They are kinetic forms whose subsequent appearance becomes elusive because of the isolation of a more stable polymorph. It is not known how and why the isolation of a more stable form blocks the formation of the earlier discovered polymorph even in locations very distant from the place where the second, more stable polymorph was made. The disappeared polymorph has sometimes reappeared many years or even decades after it originally disappeared. Sometimes, the disappeared polymorph has never reappeared despite intense experimentation. Evidence of the phenomenon of disappearing polymorphs is sometimes anecdotal but it can be a real problem in industry, especially if the earlier discovered (kinetic) polymorph is the desired one. There is also a considered opinion among researchers that the entire issue of disappearing polymorphism is an artificial construct — no polymorph can really disappear. It is just that the exact conditions for its appearance during crystallization have not been reproduced again.

Another commonly used term is *pseudopolymorphism* and it is discussed further in Section 6.4. For now, let us say that it refers to the inclusion of solvent in a crystal, more specifically to crystal forms in which an organic molecule is associated with differing amounts of a solvent molecule, for example crystals with stoichiometries like $M.H_2O$, $M.(H_2O)_{0.5}$, $M.(H_2O)_{1.5}$ and $M.(H_2O)_2$. Such hydrates are commonly found in the pharmaceutical industry, where M refers to the formula of the drug. However, pseudopolymorphs are not polymorphs in the real sense (the word *pseudo* means false) but rather correspond to a completely different chemical situation. Pseudopolymorphs (especially hydrates) are of great interest to pharmaceutical chemists because they often have different solubilities, bioavailabilities and patent rights. Certain compounds, like caffeine, have never been crystallized in an unsolvated form.

We will now discuss the prevalence of polymorphism in molecular solids. Later we develop thermodynamic relationships between polymorphs.

5.2 Occurrence of Polymorphism

There is no doubt that organic compounds give many polymorphic crystals. It is useful to ask how common this phenomenon really is. Some compounds seem to yield polymorphs readily. A few examples are naphthazarin, pyrazine-2-carboxamide, thiourea and 2-thiobarbituric acid. There are cases known where more than 10 polymorphs are known for a single compound; many of these are obtained with relatively easy experimentation. There are other compounds that have never yielded a second crystal form despite intensive experimentation over the decades. Benzoic acid, D-glucose, urea and naphthalene are examples of common compounds that have been recrystallized hundreds of thousands of times and have yet never given a polymorph under ambient pressure conditions. Then there are vast numbers of compounds in the intermediate category that will yield new crystal forms provided a lot of experimentation is carried out. Clearly there is no hard and fast rule that will predict the frequency of polymorphism in organic compounds. In the 1970s, Walter McCrone made a provocative statement that is widely quoted. He said that "every compound has different polymorphic forms and the number of forms known for a given compound is proportional to the time and energy spent in research on that compound".

Walter McCrone (1916–2002) made a very provocative and much-quoted statement regarding the occurrence of polymorphs.

Calculations show that for any given organic molecule there are a very large number of hypothetical crystal structures that are close in energy to the lowest energy structure, which is called the global minimum. Typically, one may be able to identify around 100–150 crystal structures within 2 kcal mol^{-1} from the global minimum. However, most of these structures are not realizable experimentally. Isolable polymorphs occur for compounds in which there are strong reasons for the formation of densely packed structures *and* structures in which there are specific intermolecular interactions that are chemically very favorable. Densely packed structures are generally the most stable ones. Structures with favorable directional interactions are generally formed faster. This conflict between thermodynamic and kinetic factors leads to polymorphism. We will discuss this more in Section 5.4 but at this stage it is useful to know that compounds that have never given polymorphs are the ones in which the kinetic and thermodynamic form is one and the same. The crystal form that is obtained the fastest is also the most stable.

A survey of the Cambridge Structural Database (Section 2.6.2) tells us that around 5% of the entries represent polymorphs. This figure may be an underestimate. If McCrone is correct, 100% of all compounds have polymorphic modifications, but most of them have not been found. This is probably an overestimate. A group in Innsbruck, Austria has been active in the detection of new polymorphs of organic compounds, for over 70 years. This group was headed in the past by Ludwig Kofler, Maria Kuhnert-Brandstätter and Artur Burger, and presently by Ulrich Griesser. By the estimate of this group, around a third of all organic compounds are capable of ambient pressure polymorphism. This number sounds reasonable, and will probably be accepted by many researchers in the crystal engineering area.

Ludwig Kofler (1891–1951) with the hot stage microscope made famous by him for the detection of new polymorphs, the measurement of their thermal properties and interconversions. See also Section 5.5.

5.2.1   *Polymorphism and Intermolecular Interactions*

It would always be useful to be able to predict if a particular compound would give polymorphs or not. Do certain functional groups tend to result in more polymorphs for a compound? We have already mentioned flexible groups that can also form hydrogen bonds as partly responsible for polymorphism in drug molecules. An analysis in terms of hydrogen bonded supramolecular synthons (Section 3.3) is instructive, with O–H···O, N–H···O, C–H···O and C–H···N interactions being implicated the most often.

Three distinct situations are worthy of comment in polymorphic crystals: (i) the same synthons are formed by the same functional groups in the polymorphs but the differences in the overall packing are caused by variations in the rest of the packing; (ii) the same synthons are formed by the same functional groups but there are multiple occurrences of these groups in different and distinctive molecular locations leading to polymorphism; (iii) different synthons are formed, leading to radically different packing arrangements in the polymorphs. Let us illustrate each of these situations, in turn.

The first situation is nicely exemplified in the dimorphs of resorcinol (refer adjacent diagram). Here, the hydroxy groups form the O–H···O–H synthon. This synthon can, in general, be extended to trimers, tetramers, rings and infinite chains. In resorcinol, it is elaborated in two alternative ways. In the form obtained at low temperature, A, the hydrogen bonds form tetramer helices. In the high temperature form, B, the hydrogen bonds form linear infinite chains. Resorcinol is unusual in that the high temperature form is the less dense polymorph. This anomalous occurrence owes to the open hydrogen bond network in the high temperature form.

The second situation, in which the same synthon is formed by the same functional groups in distinct molecular locations, is illustrated by the well known anti-tuberculosis drug pyrazinamide (refer diagram). This compound has four polymorphs. The amide dimer synthon is the major hydrogen bonded pattern. The minor pattern is a C–H···N dimer formed between the aromatic C–H groups and the heterocyclic N-atoms. But there are three distinct aromatic C–H groups located in different locations in the molecule (positions 3, 5 and 6 counted from the heterocyclic N-atom adjacent to the amide group). Depending on which H-atom we choose to form the C–H···N dimer synthon, there are three possibilities, and each of them occurs in one of the polymorphs. The fourth polymorph has a quite different packing but even here, a distorted version of the C–H···N dimer is found.

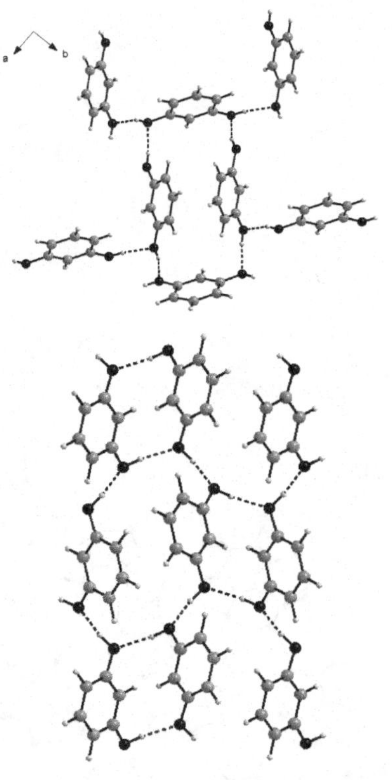

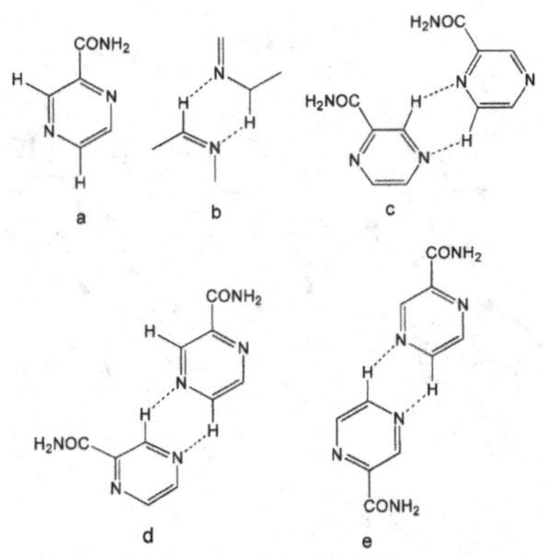

Dimorphic resorcinol: The low temperature Form A (top) shows tetranuclear patterns. The high temperature Form B (bottom) shows linear chains. Notice that the hydroxyl group conformations in the two morphs are different.

C–H···N hydrogen-bonded supramolecular synthon (b) of pyrazinecarboxamide (a) is optimized in three polymorphs with alternative arrangements c through e.

Finally, we have many interesting crystal structures of polymorphs in which the packing and synthons are quite different in the various forms. Let us consider 2,6-dihydroxybenzoic acid, which is one of several hydroxybenzoic acids in which O–H···O interactions play a significant role. Generally, carboxylic acids form symmetrical hydrogen bonded dimers (Section 3.3.2). The picture changes when additional OH groups are present especially when intramolecular hydrogen bonding is possible. In form A of this acid, the carboxyl hydrogen atom is *intra*molecularly hydrogen bonded to an oxygen atom of a neighboring OH group, the hydrogen atom of which is *inter*molecularly hydrogen bonded to the carbonyl oxygen atom of another carboxyl group. The hydrogen bond pattern in form B is completely different. The acid group forms the more conventional dimer arrangement and both the ortho OH groups form intramolecular hydrogen bonds with the two oxygen atoms of the carboxylic acid group.

5.3   Thermodynamics of Polymorphism

Crystallization of polymorphs is difficult to control as the outcome is kinetically complicated by competitive nucleation and crystal growth processes. In Chapter 4, we showed how the nucleation and growth processes are related to thermodynamic properties such as solubility, interfacial energy and supersaturation; these properties are different for each polymorph. The relative stability of a polymorph depends on its free energy: a more stable polymorph has a lower free energy. In a defined set of temperature and pressure conditions, the most stable polymorph is called the thermodynamic form (global energy minimum) while the other polymorph or polymorphs are termed kinetic or metastable forms (local minima). A metastable polymorph is thermodynamically unstable but it has a finite existence whose duration depends on its rate of transformation to more stable forms. Sometimes, the rate of transformation is infinitesimally slow and a kinetic polymorph becomes indefinitely stable; diamond never converts into graphite spontaneously. The extent of polymorphic transition depends on the crystallizing conditions and also on stabilities of the various polymorphs. Thermodynamics provides us

(a)

(b)

Synthon polymorphism in 2,6-dihydroxy-benzoic acid: (a) Form A; (b) Form B.

with a framework within which the relative stabilities of the polymorphs can be assessed.

Crystallization of a compound is generally carried out from a solution, melt or vapor (Section 4.1). In selected cases, crystallization of a new phase can also be induced in the solid state by external stimuli like heat (T) or pressure (*p*). We shall assume that the solid state crystallization does not proceed through melting, sublimation or vaporization. In other words, the equilibrium occurs strictly between two solids.

When a compound exists in various solid state forms or polymorphs, two major issues need to be addressed: (i) The relative stabilities of the solid phases and the transformations that can occur between the forms; (ii) The time needed for the transformations to reach equilibrium. Thermodynamics provides information about the first aspect (how far) and kinetics about the second (how fast). In many cases, a particular polymorph is metastable, meaning that it can eventually transform into a more stable state. In some systems the phase transformations are relatively rapid while in others they are infinitely slow (diamond → graphite). The transition from one polymorph to another usually occurs most rapidly when the crystals are suspended in solution. However, many materials will undergo transformation even in the solid state. Solution-mediated phase transformations proceed faster the higher the solubility and the greater the solubility difference between the two forms.

5.3.1 *Free Energy Diagrams and Stability of Polymorphs*

The relative thermodynamic stability of solids and the driving force for a transformation at constant temperature and pressure are determined by the difference in Gibbs free energy, $\Delta G = \Delta H - T\Delta S$. While the enthalpy difference between the forms, ΔH, reflects the lattice or structural energy differences, the entropy difference, ΔS, is related to the disorder and lattice vibrations. The relative stability is given by the sign of ΔG. When the free energy decreases, ΔG is negative and the transformation can occur spontaneously; when $\Delta G = 0$, the solid phases are at equilibrium with respect to the transformation; when ΔG is positive, the free energy increases and the transformation is not possible. The thermodynamic condition for equilibrium between different phases and the possible directions for the transformations at constant

This unusually shaped rock formation looks as if it should collapse, but this never happens. Similarly, a kinetic polymorph is not the most stable form but it can remain indefinitely without converting into the thermodynamic form.

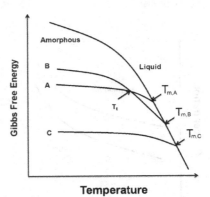

Temperature

Gibbs free energy versus temperature curves for three polymorphs A, B, and C with melting points at $T_{m,A}$, $T_{m,B}$ and $T_{m,C}$. C is the thermodynamically stable crystal while A and B are kinetic forms. T_t corresponds to the transition between B and A. The amorphous solid has a higher free energy than the crystalline forms. A undergoes a phase transition to B. A and C (or B and C) are monotropic pairs while A and B are enantiotropes.

The n-phase one component system is enantiotropic if there are a series of (n−1) phase transformations below the melting point, and monotropic if there are no such transformations. This assumes that the transition involved is at a fixed pressure (generally atmospheric pressure).

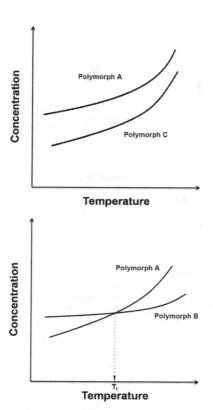

A plot of concentration versus temperature for a monotropic pair of crystals A and C (top), and for an enantiotropic pair A and B (bottom).

pressure for a single component system can be illustrated using Gibbs free energy (G) versus temperature (T) diagram. Let us consider a system containing three polymorphs A, B and C in which C is the thermodynamically stable phase. The melting points of A, B and C respectively are $T_{m,A}$, $T_{m,B}$ and $T_{m,C}$. C is more stable than A or B since $\Delta G < 0$ for the A → C as well as B → C transformations. In the G versus T diagram, the intersection points represent phases that co-exist in equilibrium. At the melting temperature, the liquid exists in equilibrium with a solid while transition temperatures refer to the transformation of one polymorph to another.

5.3.2 *Monotropes and Enantiotropes*

A pair of polymorphs is classified as *monotropic* (say forms C and A in the G versus T diagram in the adjacent box) where one form (C) is more stable than the other at all temperatures below melting. The forms are referred as *enantiotropic* (forms A and B in the diagram) where there is a transition temperature (T_t) below melting. Above and below the transition temperature, the stability order is reversed for enantiotropes. Each enantiotrope has its own temperature range of stability. However, one monotrope (for example C) is always more stable than the other under all conditions in which the solid can exist. The transition temperature is defined as the temperature at which the free energy isobar of polymorph A intersects the free energy isobar of polymorph B. Thus, at T_t both polymorphs have equal free energy, i.e., $G_A = G_B$ and consequently are in equilibrium with each other. Below T_t, polymorph A is more stable solid phase because $G_A < G_B$. Below T_t, B can undergo a spontaneous exothermic transformation into A. Above T_t, B is more stable because its free energy is lower than that of A, or $G_B < G_A$. Therefore, above T_t and below melting, A can undergo a spontaneous exothermic transition to B.

The Gibbs free energy, ΔG, and its temperature dependence can be obtained from melting data (melting temperature, enthalpy of fusion), enthalpy of transformation if available, and/or solubility dependence on temperature data for the solid phases of interest (enthalpies of solution from van't Hoff plots). The melting point method requires a smaller amount of sample. The solubility method is advantageous to investigate a range of temperatures and various solvents. Also, there is the possibility of solution mediated transformations and/or solvate formation during the solubility measurements.

The thermodynamic behavior of monotropes and enantiotropes can also be described using solubility versus temperature diagrams (refer to the adjoining diagram). The Gibbs free energy difference between two polymorphs reflects the ratio of their activities. The activities can be approximated by the solubilities; the high energy polymorph will have a higher thermodynamic activity or solubility. Similarly it can be shown that the high energy polymorph will have a higher vapor pressure, dissolution rate or rate of reaction. In the monotropic case, the stable polymorph will have the lowest solubility. In the enantiotropic case, the solubility will be reversed at the transition point.

Generally, solid to solid transformations are kinetically hindered because of the activation energy associated with them. In many cases, solid–solid transformation occurs at a temperature that provides the system with sufficient thermal energy to cross the activation energy barrier. The transition point (T_t) in a monotropic system is a *virtual transition point*, because it lies above the melting points of both polymorphs. This notion assumes that the free energy curves of the two polymorphs converge beyond their melting point. Artur Burger and Rudolf Ramberger formulated the heat of transition rule and the heat of fusion rule to determine whether the relationship between a pair of polymorphs is enantiotropic or monotropic.

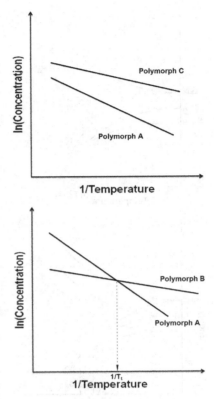

A plot of ln(concentration) versus temperature for a pair of monotropes A and C (top) and enantiotropes A and B (bottom).

5.3.2.1 *Burger-Ramberger Rules*

The enthalpy of transition rule states that if an exotherm is observed, no transition lies below that temperature. If an endotherm is observed, then the transition point must be at, or below, the observed endotherm. The enthalpy of fusion (melting) rule states that if the higher melting of the two polymorphs has the lower enthalpy of fusion, then the relationship is enantiotropic. Conversely, if the higher melting of the two polymorphs has the higher ΔH_f°, then the relationship is monotropic.

Burger's heat of transformation rule can be ascertained by concentrating on the H curves and seeing what happens on going from H_A and H_B and vice versa, remembering that this is only possible by lowering the free energy, i.e. ΔG must be positive. Hence,

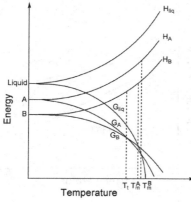

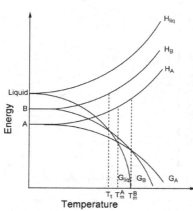

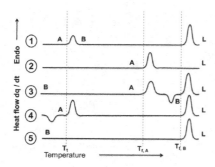

those processes which are exothermic on raising the temperature are spontaneous ones. Spontaneous processes are necessarily irreversible, so this transition will be irreversible at or below that temperature. The reverse applies to endothermic processes. Burger's heat of fusion rule depends on H_A and H_B being approximately parallel so that the difference in heat capacity (C_p) does not obscure the differences in the heats of transition.

Burger's rules are for the interpretation of the experimental observations, not a substitute for them. The heat of fusion rule should be used only if the transition cannot be measured.

5.3.2.2 *Distinguishing between Enantiotropes and Monotropes*

Differential scanning calorimetry (DSC) or differential thermal analysis (DTA) scans (Section 5.5.3) can be employed to distinguish an enantiotrope from a monotrope. Let us consider the same system considered above. Let us examine the qualitative DSC scans for an enantiotropic system, for example A and B. Scans 1 through 3 correspond to the heating of B while scans 4 and 5 correspond to the heating of A. Scan 1 shows two endothermic transitions, the first corresponding to a phase transition from A to B followed by the melting of B. Scan 2 shows only one endothermic transition corresponding to the melting of B. Here, the phase transition between B and A is probably not observed due to kinetic reasons (e.g. fast heating rate). Scan 3 shows the occurrence of two endothermic and one exothermic peak. The two endothermic peaks correspond to the melting of A and B respectively. The exotherm represents the phase transition from A to B; B crystallizes from the melt at a higher temperature. Scan 4 shows three peaks: the first exothermic peak corresponds to the transformation of the metastable B into stable A and again to B which is the stable phase at higher temperatures and hence endothermic. The third peak corresponds to the melting point of B. Scan 5 shows only one transition due to the melting of B. The other transitions are not observable due to kinetic reasons like a fast scan rate. DSC/DTA runs accompanied by thermogravimetry (TG) would be helpful in eliminating the possibility of weight loss due to decomposition. See Section 5.5.3.

5.4 Thermodynamics versus Kinetics and the Formation of Polymorphs

Thermodynamics establishes the stability of the various equilibrium phases. However, the kinetic pathways determine which solid will be formed and for how long it can remain stable. Let us consider the crystallization process in terms of molecular events. Initially the molecules arrange themselves into energetically suitable aggregates or packing patterns under the action of non-covalent forces, notably hydrogen bonds and other directional interactions. The balance between the kinetic and thermodynamic factors can be illustrated by the free energy reaction coordinate (Section 4.3) for a transition from the initial state G_i, to two different polymorphs C and A. Assume that C is more stable and less soluble than A ($G_C < G_A$). G_i may represent a supersaturated solution in a multiple-component system, liquid or solid (molecular dispersion in amorphous system), or in the case of a single component system a supercooled liquid (melt), or an amorphous solid. The reaction follows a path through an energy maximum between the initial and final states. This resistance to the transition from G_i to G_C or G_A arises because there is an activation energy for molecular diffusion, molecular assemblies, and for the creation of an interface. For a chemical reaction in a homogeneous system, this energy maximum is the transition state and reflects elementary reactions, bimolecular or trimolecular, that yield products with new covalent bonds. In comparison, a crystallization process or phase transformation leads to heterogeneous systems in which a separate new phase is created from a supramolecular assembly by formation of non-covalent bonds. A transition from the initial state G_i to the phase C or A will depend on the activation energy. According to the reaction pathway, the energy barrier for C ($G_C^* - G_i$) is greater than that for A ($G_A^* - G_i$). Because the rate of nucleation is related to the activation energy on the reaction path, A will nucleate at a faster rate than C even though the change in free energy is greater for C ($G_C - G_i$) than for A ($G_A - G_i$). This behavior of polymorphs appearing in the order of increasing stability is referred to as Ostwald's rule of stages. This rule states that *when leaving an unstable state, a system does not seek out the most stable state, but rather the nearest metastable state which can be reached with least loss*

All chemical reactions are subject to the dictates of thermodynamics and kinetics — how far and how fast? Most chemical reactions wherein covalent bonds are made and broken are kinetically controlled. This is also true of crystallization which is a supramolecular reaction. Like other kinetic processes, the outcome of crystallization depends a great deal on temperature, solvent, rates of heating and cooling, impurities, shock, and other experimental variables. A particular crystal form may not be the global minimum in free energy but it may be obtained repeatedly because it is kinetically locked in under the conditions employed. The appearance of these local minimum structures during crystallization is one of the main reasons for polymorphism. That kinetic products may dominate or even be formed exclusively in chemical reactions was shown by David Y. Curtin and Louis P. Hammett, in the 1950s. The Curtin-Hammett principle states that the distribution of products in a reaction that has many pathways need bear no relation to the relative stabilities of these products. Exactly the same holds for crystallization. Given a collection of molecules that can come together in many ways, the favored route has little to do with the stability of the final ensemble but rather with how fast this route can be travelled.

of free energy. However, Ostwald's rule of stages is not universally valid because the appearance and evolution of solid phases are determined by the kinetics of nucleation and growth under the specific experimental conditions and by the links between molecular assemblies and crystal structure.

We have already shown in Chapter 4 that crystallization involves both the nucleation and growth of a solid. Nucleation plays a key role in the selective crystallization of polymorphs and the stabilization of metastable states. Studies of growth kinetics and crystal morphologies are useful in characterizing intermolecular interactions on specific crystal planes and as a consequence in identifying additives or solvents that may either promote the growth of a particular polymorph or preclude the nucleation of a solid by forming solvates or co-crystals (refer Section 4.6).

5.5   Methods of Polymorph Characterization

Methods for characterization of organic crystals have been summarized in Chapter 2. Some of these and also other methods are particularly appropriate to characterize and distinguish polymorphs. The following are currently the most popular.

5.5.1   *Hot Stage Microscopy*

Hot stage microscopy or thermal microscopy is a useful technique for studying polymorphism. The method provides insights into the thermal transitions in polymorphs. The method involves viewing the sample using some form of optical microscopy while also employing a hot stage to control the temperature of the sample. In addition to measuring transition temperature, the information derived can be broadly used to identify changes in crystal habit associated with the polymorphic change and identifying changes in light transmission or reflection associated with the transition.

5.5.2   *X-ray Diffraction*

X-ray diffraction (XRD) is the most important tool for polymorph identification and quantification. A distinction can be made between two main XRD techniques: the X-ray beam can be focused onto a single crystal or

Hot stage micrograph of the 1:2 co-crystal of 4,4'-bipyridine and 4-hydroxy-benzoic acid at 195 °C shows the growth of a more stable polymorph from the surface of a crystal of a less stable form.

on powder spread on a sample holder (PXRD). In both techniques, the diffractogram thus obtained can be considered as the *fingerprint* of a substance. It can provide information about the phases present (peak positions), phase concentrations (peak areas), and the amorphous content (background hump). In the study of pharmaceutical polymorphs, it is often not possible to grow sufficiently large single crystals suitable for single crystal XRD. Generally, if two specimens have the same powder X-ray traces, they are considered to represent the same crystal form. By *same* we mean powder traces where the same lines are present (with 2θ values ± 0.1 to $0.2°$ in the two cases) and intensity variations that are no more than 20% in the two cases.

5.5.3 *Thermal Analysis*

Differential scanning calorimetry (DSC) is an important technique for the study of polymorphs, providing a means both for identification and for characterization. DSC works on the principle of power compensation, whereby the sample and reference are maintained at the same temperature and the heat flow required to keep the two at thermal equilibrium is measured. This allows for the measurement of temperature and the energy associated with a thermal event to be easily assessed. The energy associated with the transition is calculated using $dQ/dt = \Delta T/R$, where dQ/dt is the heat flow, ΔT is the temperature difference between the furnace and the crucible and R is the thermal resistance of the heat path between the furnace and the crucible. DSC provides raw data in the form of heat flow (power in mW) plotted against temperature (K), the former referring to the heat flux difference between the sample and reference. The basic parameters obtained from the measurement indicates that the power ($J s^{-1}$) at a constant scanning rate ($K s^{-1}$) will be related to the heat capacity via $dQ/dt = C_p dT/dt$ where dQ/dt is the heat flow and dT/dt is the heating rate. In effect the measurement gives the heat capacity of a material. As the sample undergoes a thermal event it is effectively altering the total heat capacity of the system due to the latent heat associated with the melting, crystallization etc., this being seen as a peak or, in the case of a glass transition, a shift in the baseline. Integration of the above equation indicates that the area under the curve for a thermal event will be proportional to the energy involved in the process. Hence, by suitable calibration with standards with known melting points, heats of fusion and heat capacities, it is possible to characterize a range of samples. DSC is used to differentiate between polymorphs on the basis of melting point but may also be used to study the transformation behavior of metastable systems and to probe the nature of the interrelationship between the polymorphic forms.

Differential thermal analysis (DTA) is another technique similar to DSC. In DTA the sample and the reference are heated at a constant heating rate at controlled atmosphere; the difference in temperature (ΔT) between the sample and reference is plotted against temperature. Endothermic changes (melting, evaporation) are shown as negative and exothermic changes (decomposition) are shown as positive. Although this technique is useful, it is not as quantitative as DSC.

Thermogravimetric analysis (TGA) is a technique which allows the monitoring weight change as a function of temperature. Unlike DSC, this is not useful for the study of polymorphs as the transitions does not involve weight change. It is useful to characterize solvates, hydrates and host-guest solids, in quantifying their stoichiometry and assessing their stability.

5.6   Properties of Polymorphs

Polymorphs have different crystal structures. Therefore, properties that depend on the internal structure of a crystal will be sensitive to the polymorph that is being considered. Polymorphism in molecular crystals provides an opportunity to study structure-property relationships. A historically important example of polymorphs that show different photochemical properties is 2-ethoxycinnamic acid (Section 1.2). This compound exists in three polymorphs: the α-form reacts photochemically to give a centrosymmetric α-truxillic acid dimer; the β-form reacts also photochemically but the product is the β-truxinic dimer; the γ-form is photostable.

5.6.1   *Color*

It is not often that different polymorphs have different colors but when such an event occurs, the detection and separation of polymorphs becomes quite easy. A dramatic example is furnished with 2-(4-anisyl)-1,4-benzoquinone, for which red and

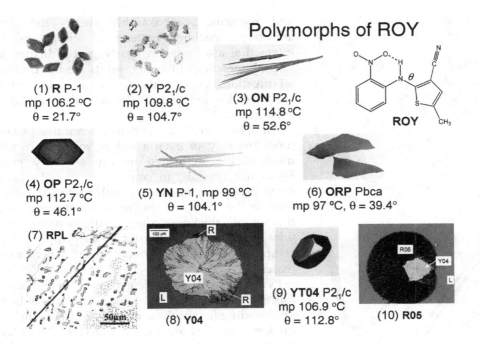

Polymorphs of ROY

(1) **R** P-1
mp 106.2 °C
θ = 21.7°

(2) **Y** P2₁/c
mp 109.8 °C
θ = 104.7°

(3) **ON** P2₁/c
mp 114.8 °C
θ = 52.6°

ROY

(4) **OP** P2₁/c
mp 112.7 °C
θ = 46.1°

(5) **YN** P-1, mp 99 °C
θ = 104.1°

(6) **ORP** Pbca
mp 97 °C, θ = 39.4°

(7) **RPL**

(8) **Y04**

(9) **YT04** P2₁/c
mp 106.9 °C
θ = 112.8°

(10) **R05**

yellow crystals are obtained concomitantly from hexane solution. The molecule consists of donor and acceptor moieties, D and A. These are, respectively, the anisyl ring and the quinonoid ring. In the yellow form, the molecules are stacked DDDD.. AAAA.. as is seen in the crystal structures of many other quinones. In the red form, the donors and acceptors alternate in the stack DADADA... The red color of the solid is on account of the charge transfer interactions that arise from the overlapped stacking of D and A molecules.

A unique compound that shows many polymorphs in differing colors and morphologies is 5-methyl-2-[(2-nitrophenyl)amino]-3-thiophenecarbonitrile, or ROY, a compound that has no less than seven polymorphs: red prisms, orange needles, yellow needles, orange-red plates, orange plates, and two types of yellow prisms. Why a single compound should have so many polymorphs is intriguing but the reason must have to do with a fortuitous combination of kinetic and thermodynamic factors.

5.6.2 *Mechanical Properties*

Organic crystals such as hexachlorobenzene, hexabromobenzene, 1,3,5-trichlorobenzene, 1,3,5-tribromobenzene

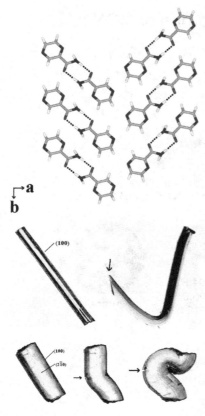

Bending crystals can be deformed into many shapes and in some cases they can even be flattened onto themselves without breakage. They are highly plastic and ductile. In contrast, crystals with comparable intermolecular interactions in all three directions do not show bending and will be hard and brittle, whether the interactions in themselves are strong (hydrogen bonds) or weak (van der Waals). In these isotropic crystals, the packing patterns are similar in the three orthogonal directions and highly cross-linked structures (naphthalene, urea, benzoic acid, D-glucose). Cross-linking in the context of aromatic molecules is equivalent to herringbone packing.

Mechanical properties of molecular crystals are important for large-scale processing and handling of materials in industry, especially the pharmaceutical industry. Soft crystals become pasty upon grinding. Harder crystals are granular and can be handled more easily.

and pyrazine-2-carboxamide show characteristic anisotropic packing; patterns of strong and weak interaction are found in nearly perpendicular directions. Interactions in such crystals are not uniform in all directions; the crystals are not isotropic. Such crystals can be bent, and the bending is strongly anisotropic.

An organic crystal that bends is not like plasticine, modeling clay, or even a slab of metal. It cannot be mechanically deformed in any arbitrary manner. Molecular crystals, in contrast to metallic crystals, show practically no change with respect to their volume, in the lengths of the inner and the outer arcs or the sample thickness after plastic bending. This is why the shape of the plastically deformed molecular crystal appears to be different from that of metallic materials.

Pyrazine-2-carboxamide exists in four polymorphic forms and crystals of the α modification can be bent. The bending face of these crystals is (100) which is the thickest (smallest area) crystal face. The thin face of the crystal is not the bending face. The amide dimers are stacked along the short axis [001] at a distance of 3.72 Å; there are only weak N–H···N (2.46 Å, 135°), C–H···O (2.49 Å, 167°) and C–H···N (2.55 Å, 175°) interactions present in the two other orthogonal directions. In this crystal, one set of interactions (stacking) is significantly stronger than another (C–H···N) in a perpendicular direction. As a consequence of bending, the angles between some of the faces change markedly. In the bending process, cracking always starts in the region of maximum tensile stress. This example shows that bending is a consequence of crystal packing; it does not depend on anisotropy of morphology. It is not the thinnest face that always bends.

5.6.3   *Chemical Reactivity*

The reactivity of molecules in the solid state depends substantially on the way in which the reactive functional groups are arranged in the crystal and/or on how the functional groups are exposed and accessible to external molecules. Since the internal packing arrangement is not the same in polymorphic crystals, differences in the rate of reactions of polymorphs and in their thermal or photochemical decompositions need not be surprising.

5.6.3.1 *Polymorphism in Energetic Materials*

High energy materials can be defined as those that release heat and/or gaseous products very rapidly when stimulated by heat, impact and shock or sparking. These substances may be broadly classified as explosives, propellants, and pyrotechnics. The performance of high energy materials depends on their sensitivity to detonation by external stimuli, detonation velocity, thermal stability, crystal density, chemical reactivity and crystal morphology; these properties in turn depend on the polymorph that is being considered.

The commonly used secondary explosive HMX (1,3,5,7-tetranitro-1,3,5,7-tetraazacyclooctane, octogen or cyclotetramethylenetetranitramine) exists in four crystalline forms (the γ-form is a hydrated form). Of these, the β-form is easy to handle since it has the lowest sensitivity to impact (shock). Therefore, β-HMX is preferred for practical applications. The detonation velocity of a high energy material is proportional to its density. Hence it is not surprising to see that the densest polymorph of an energetic compound is usually selected. The crystal morphology of a given polymorph is important in the processing and packaging of explosives. For example, prismatic crystals pack more efficiently than crystals with rod, needle, or platelet morphologies.

Hexanitrohexaazaisowurtzitane (HNIW) or CL-20 is another interesting explosive which was developed as a propellant. It is one of the densest and most energetic organic compounds known at present. Under ambient conditions, four forms are known of which the α-form is a hemihydrate. The stability sequence is ε > γ > α > β but the order of the densities is γ (1.916 g cm^{-3}) < β (1.985 g cm^{-3}) < α (2.001 g cm^{-3}) < ε (2.04 g cm^{-3}). This indicates that the most densely packed structure need not be the most stable crystal. Such an occurrence is uncommon but not completely unknown.

5.6.3.2 *Polymorphism and Reactivity of Drugs*

Drug formulations are usually shelved at ambient temperature. Their reactivities towards water vapor and air and their thermal stabilities affect the shelf life of the tablets despite the fact that the drugs are usually encapsulated inside a polymeric coating. The crystal

HMX

HN1W

Phase transitions in high energy solids, during processing or storage, may create surface and internal defects and the particle size may be reduced. It is easier to initiate explosive reactions at these defect sites. Therefore the sensitivity of an energetic material to initiation is related to the number of defects in a crystal. Shear stresses in the solid can also contribute to the initiation processes.

Indometacin

Metastable polymorph

A metastable polymorph of a drug is sometimes desirable because of special properties such as higher bioavailability, behavior during grinding and compression or lower hygroscopicity. However, a metastable form has a thermodynamic tendency to reduce its free energy by transforming into the stable form. Such a polymorphic transformation is often detrimental to the efficacy of the formulation. Examples include chloramphenicol palmitate and ritonavir. Further, manufacturing processes such as compaction, milling, wet granulation, and freeze drying can also result in polymorphic transitions.

structure can affect the reactivity of a drug molecule in the solid state.

Indometacin or indomethacin is a non-steroidal anti-inflammatory drug commonly used to reduce fever, pain, stiffness, and swelling. It exists in three modifications. The amorphous form readily reacts with ammonia to give the corresponding amorphous ammonium salt while the α-form produces a microcrystalline salt. The stable γ-form is, however, inert to gaseous ammonia. The reactivity of amorphous indomethacin is due to the higher molecular mobility and free volume. The differences in reactivity between the α- and γ-forms can be attributed to the arrangement of the molecules within the respective crystal structures.

5.7   Case Studies from the Pharmaceutical Industry

We now give details as to how polymorphism became a significant issue in the study of some important pharmaceuticals. Polymorphism is of unusual importance in the pharmaceutical context. In addition to pure chemical compounds, polymorphs are also formed by co-crystals, solvates and salts. The phenomenon of polymorphism extends therefore to all structural variations around a drug molecule, and indeed it defines the *structural landscape* of the compound in question.

5.7.1   *Ranitidine*

This block buster anti-ulcer drug, in the form of the hydrochloride of the free base, was developed by Glaxo, and the first patent (called the '658 patent) was issued in 1977 for the drug. During the scale-up work, it was found that a sample was obtained that gave a different X-ray powder diffractogram from the others. Glaxo concluded that they had obtained a new polymorph, which they called Form 2. The material that was obtained according to the procedures outlined in the '658 patent was thereafter referred to as Form 1. In 1985, Glaxo was granted a patent on Form 2, and this was referred to as the '431 patent. Glaxo marketed ranitidine hydrochloride under the brand name of Zantac.

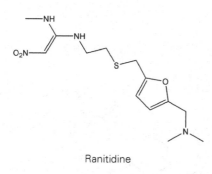

Ranitidine

By 1991, sales of Zantac reached $3.5 billion and this single drug was practically the lynchpin of the entire Glaxo group. Several generic companies prepared to go on the market with Form 1 in anticipation of the expiration of the '658 patent in 1995. One of these companies, Novopharm, was notable for the long and intense legal battle it undertook with Glaxo. Initially Novopharm, which could not make Form 1 according to the '658 patent, claimed that '658 inevitably, produced Form 2. It therefore claimed that the '431 patent was, in essence, invalid and it sought to market Form 2. Glaxo successfully defended its patents and proved that Form 1 could indeed be made using the procedures of '658. Novopharm lost to Glaxo in the first round of litigation.

However, by 1994, Novopharm had developed a method of making Form 1, practically free of Form 2 (less than 1%), and sought to market the same. Glaxo now sued Novopharm and claimed that Novopharm's samples of Form 1 contained Form 2, which was still protected by '431. However, Novopharm showed that its samples of Form 1 did not contain detectable amounts of Form 2. The court stated that if indeed Novopharm's samples of Form 1 contained a little of Form 2 (≤1%), then this impurity was just that, an impurity, and not the basis of some improvement of the drug. Therefore Novopharm was allowed to market mixtures of Form 1 and Form 2.

This example was the first major legal event that brought the scientific field of organic polymorphism into the glare of commercial and industrial activities. Ever since then, pharmaceutical companies have been aware of the importance of crystal engineering to their activities. This example also showed the need for increasingly better equipment that can detect ever smaller amounts of infringing polymorphs. This is known as limit of detection.

5.7.2 *Ritonavir*

Ritonavir is an important peptidomimetic AIDS drug introduced in 1996 and sold by Abbott Laboratories under the trade name Norvir. Originally, the drug was dispensed as an ordinary capsule, which did not require refrigeration. It was not known at the time the drug was first marketed that the form in which it was being marketed is a kinetic form. The crystal description of this form is in the monoclinic system and it was subsequently referred to as Form I.

As a matter of convention, a patent is referred to by its last three numbers. So patent 4128658 for ranitidine is referred to as the '658 patent.

Ritonavir

In 1998, a second polymorph (Form II) appeared in one of the production batches of the drug. The second form was discovered as a result of the observation that several lots of Norvir capsules began failing dissolution specifications. Evaluation of the failed drug products revealed that a second crystal form had precipitated from the formulation. This second modification is a conformational polymorph. It is also the thermodynamically stable form. None of these facts is in itself a complicating factor. However, there were two additional factors that pertained to Form II, and these factors quickly contributed to a major crisis for Abbott that completely upset the production of the marketable form of the drug. The first of these factors was that the lower energy second polymorph caused the therapeutically effective higher energy polymorph to convert to the lower energy polymorph on contact. The other factor was that the second form was pharmaceutically ineffective because of its much lower solubility (only 50% that of Form I). Very quickly, every batch of the desired polymorph was converted into the undesired Form II, and the kinetic polymorph Form I could also not be obtained.

Abbott was then forced to remove the oral capsule formulation of the kinetic form from the market and this had serious implications for the marketed product and also for the patients taking the drug. In the academic world, the first form of ritonavir was much quoted as an example of a disappearing polymorph (Section 5.1.1). Abbott put in a considerable effort and spent much money in trying to obtain Form I again. These measures included taking stringent precautions that might suppress the accidental nucleation of Form II in the production facilities, and went all the way to building new production facilities in distant corners of the world wherein it was believed that nuclei of the new form had not yet entered! Unfortunately, none of these measures were really effective.

After some time, it began to be clear to the company that Form I, the therapeutically effective polymorph, was not going to be obtained so easily. Substantial effort went into understanding the nature of the polymorphic transformation, reformulating the drug, and identifying strategies to regenerate the original Form I. At some considerable cost a new formulation of Norvir was finally developed and launched as refrigerated filled gel capsules. The new formulation is much more expensive than the original product.

Subsequently, *high throughput crystallization* screens uncovered more polymorphs of ritonavir. These included the two known forms and three previously unknown forms. The novel forms included a metastable polymorph, a hydrate phase, and a formamide solvate. The solvate could be converted to Form I via the hydrate phase by using a simple washing procedure, providing an unusual route to prepare the previously disappearing Form I. Crystals of Form I prepared by using this method retained the small needle morphology of the solvate and thus offer a potential strategy for particle size and morphology control.

The ritonavir episode illustrates the need for early and comprehensive identification of solid-form diversity of drug compounds. We have already mentioned that it is very important for both innovator and generic companies to be aware of all possible polymorphs and variations (solvates) of a drug. The reasons given for this in Section 5.1.1 are commercial and legal. Here, we have outlined a scientific reason. It is important for industry to know about new and undiscovered polymorphs of a drug to safeguard against the unfortunate situation that prevailed for ritonavir — Abbott was marketing a kinetic polymorph that constituted a part of an enantiotropic system, and this is always a risky and unsafe enterprise. There was not much appreciation for such matters in the mid 1990s when Norvir was first marketed. It is unlikely that a drug company would take such a major marketing decision today without being in possession of all the important scientific facts concerning the crystal forms of the compound in question.

5.7.3 *Aspirin*

For 40 years following the first determination of its crystal structure in 1964, aspirin or acetylsalicylic acid appeared to be monomorphic. Although claims of a second crystalline form were made in the literature, these were never substantiated by X-ray diffraction data and reports of differing dissolution behavior were ascribed subsequently to differences in crystal habit. In 2004, computational results described the structure of a new *in silico* crystal form that was thermodynamically viable. Shortly after, it was reported that this predicted form had been realized, apparently as a serendipitous event when aspirin was crystallized from an acetonitrile solution that also contained the

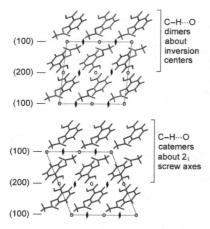

(100) —

(200) —

(100) —

C–H···O dimers about inversion centers

(100) —

(200) —

(100) —

C–H···O catemers about 2₁ screw axes

The distinction between the reported aspirin crystal structures is slight. Both contain centrosymmetric molecular dimers, held together by O–H···O hydrogen bonds between their carboxyl groups. The dimers are arranged into two-dimensional layers, which are the same in both the Form I (top) and Form II (bottom) structures. The distinction between the two structures lies in the way that these layers are stacked. In terms of the intermolecular interactions, molecules in direct contact across the layer boundary approach each other so that the methyl groups of their acetyl substituents form C–H···O contacts with the carbonyl O-atoms of neighboring acetyl substituents. In the Form I structure, molecules in direct contact define centrosymmetric C–H···O pairs, while in the Form II structure they define chain-like C–H···O motifs. The two interlayer arrangements are isoenergetic.

drug levetiracetam. Although the crystallographic characterization that formed the basis of this report was less than optimal, its appearance in the academic literature established aspirin as being polymorphic. The new form was labeled Form II, with the accompanying Form I label applied to the well-established structure. Subsequently, the new Form II was unambiguously characterized and today, crystalline forms of aspirin have been obtained that are pure Form I, pure Form II or intergrowths of the two forms. The novelty in the case of aspirin are the intergrowth forms, in which two different crystal structures are present in the same crystal, intergrown at a molecular level.

The close similarity between the Form I and Form II structures has some interesting crystallographic consequences. The existence of identical layers in the two structures, together with their specific relative translation, results in an unusual situation in which 50% of the Form I and Form II diffraction patterns are the same. Thus, 50% of the Form I and Form II structures are identical. The two structures are distinguished by the remaining 50% of their diffraction patterns, which reveals either the AAAA or ABAB stacking sequence.

The distinguishing feature of aspirin polymorphism is that one is faced with the question of what one should call the intergrown aspirin crystals. McCrone's definition of polymorphism states that *a polymorph is a solid crystalline phase of a given compound resulting from the possibility of at least two different arrangements of the molecules of that compound in the solid state.* If we use this definition, we can describe aspirin as polymorphic since there are definitely two different arrangements of the molecules in Form I and Form II.

However, difficulties arise for the intergrown crystals. The McCrone definition of polymorphism considers each different crystal structure to be a different polymorph. Molecular crystals usually crystallize with one structure or another so that a single crystal is either one polymorph or another. For intergrown aspirin, this is not the case because there are domains of both Form I and Form II in the same crystal. A possible working approach is to take the established Form I and Form II as two reference points that define the structural extremes. The intergrown aspirin crystals have structures that lie somewhere between the two reference points. The different domains in the intergrown crystals are inherently joined, and the crystals must be viewed as a coherent whole. The intergrowths are not simply mixtures of Form I and Form II.

This issue has obvious implications in pharmaceutical chemistry because it is still not clear if the intergrown crystals would qualify for separate patent protection over and above any possible protection that may exist for pure Form II.

5.7.4   *Omeprazole*

Omeprazole is a block buster anti-ulcer drug, marketed by Astra Zeneca, and its structural chemistry has several interesting features. The polymorphs of omeprazole test the current definition of the phenomenon as occurring when the same molecule exhibits different crystal structures. As long as the molecules concerned are rigid and there are no great ambiguities in their crystal structures, the meaning of the term polymorph is uncomplicated. If not, new questions arise. Notably, how similar should the same molecules be, and how dissimilar should the different crystal structures be in order for them to qualify as polymorphs?

The tautomers of omeprazole are effectively the 5- and 6-methoxy derivatives and these have also been seen in solution. Three solid forms, A, B and C, have been patented and are characterized by their powder X-ray diffractograms. Form A is more stable than Form B. Form C is easier to prepare than either Form A or B. Single crystal X-ray studies show that form B is identical to the 6-methoxy tautomer of omeprazole. Another patent, US 6,780,880, claims that omeprazole crystals contain mixtures of the two tautomers. This is confusing because there is a patented form that seems to be a mixture of two polymorphs.

Further work showed that under different crystallizing conditions, a number of structurally distinct forms can be obtained. These forms are differentiated in terms of varying proportions of the 5-methoxy and 6-methoxy tautomer. All the crystals are isostructural and Forms A, B and C are characterized by certain ranges of tautomeric composition. Accordingly, Form A has 10% of the 5-methoxy tautomer, Form B has 15% of the 5-methoxy tautomer and Form C has 12% of this tautomer. The crystal structures of the three forms are in a continuum. They may be viewed as the same structure with a statistical distribution of two slightly different molecules, the 5- and 6-methoxy tautomers.

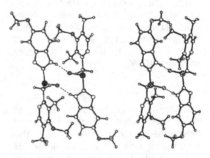

Idealized view of the 5-methoxy (right) and 6-methoxy tautomer (left) in the crystal structure of omeprazole.

Isomorphism (isostructurality) is similarity of crystal shape, unit-cell dimensions, and structure between substances of similar chemical composition. Ideally, the substances are so closely similar that they can generally form a continuous series of solid solutions. The degree of similarity between crystals can be calculated using a parameter suggested by Alajos Kalman. This is the unit cell similarity index, Π, and isostructurality index $I_i(n)$. For an isostructural pair of crystals, Π must be close to 0, and $I_i(n)$ should approach 100%.

Z denotes the number of molecules in the unit cell while Z' is the number of independent molecules in the asymmetric unit. Z' is the number of formula units in the unit cell divided by the number of independent general positions. Approximately 10% of all molecular crystals have Z' > 1. These structures are of particular interest to crystal engineers. They represent special cases in which the crystal packing is not ideal; they could arise from frustration between two or more competing factors (say optimization of hydrogen bonds versus shape packing) during the nucleation process. Systems exhibiting Z' > 1 are frequently polymorphic. Z' > 1 structures may be regarded as a special case of co-crystallization of chemically identical species. It has also been suggested that high Z' structures are representative of the fastest growing crystal nucleus, say a metastable polymorph.

However, they have different properties, notably light stability and ease of preparation.

This example of omeprazole highlights several interesting general features. The different crystal structures of the omeprazole forms are actually modulations, and the modulation is at the molecular level: they contain tautomers in differing amounts within the same crystal packing. Since the forms contain different amounts of tautomers, they are tautomeric polymorphs. But how many polymorphs of omeprazole really exist? Is it one or two or infinite? Would each tautomeric composition qualify for independent patent protection or would it be more meaningful to claim protection for compositional ranges? It is interesting to note that at present, the patented Forms (A, B and C) are defined in terms of properties (stability, ease of preparation) rather than in terms of structure. The role of the powder X-ray diffractogram is merely as a fingerprint of a form with a particular property rather than diagnostic of a particular structure type. The most significant aspect of omeprazole polymorphism is that function seems to be a more meaningful criterion of polymorph patentability than structure.

5.8   Polymorphism Today

Much of this chapter is based on differences in structure (molecular structure and crystal structure) as criteria of polymorphism. If function is more significant than structure, this raises more provocative issues: (i) Should the definition of polymorphism rely so heavily on structural differences? (ii) Are subtle structural differences really meaningful, especially in the context of the kind of modulation seen in omeprazole and also in aspirin? (iii) Just as minor differences in crystal structure may be interpreted subjectively, may this also be said of molecular sameness? (iv) Accordingly, how important is the criterion that the molecular structure should be exactly the same if two crystals are to be called polymorphs? (v) In the context of polymorphism, is it more reasonable then to speak of a structural landscape that includes a number of solvated and unsolvated variations of the same molecular species, without insisting on a rigorous stoichiometric and chemical identity? In the end, one may conclude that an all-encompassing definition of polymorphism is elusive, and also perhaps unnecessary.

The study of polymorphism is one of the most active areas of investigation in crystal engineering today. The topic is academically challenging and at the same time, industrially important. It is rare in chemistry for a topic to be very important to both academic and industrial communities *simultaneously*. New insights into polymorphism are constantly being unraveled. A few years ago, it was shown, for example, that a deuterated analog of a compound could be induced to form a new polymorph that could not be obtained for the original material. The reasons for such an enigmatic finding are still far from clear but in the end, there is little doubt that this and other similar observations will ensure that the study of polymorphs will engage the crystal engineer for several years to come.

5.9 Summary

- Polymorphs are different crystal forms of the same chemical compound. Polymorphism is a solid state phenomenon.
- Crystal engineering provides a good platform to study polymorphism because one of the concerns of this subject is an understanding of intermolecular interactions.
- Polymorphism occurs frequently in organic solids and notably in drug molecules because they contain flexible functional groups capable of hydrogen bonding. If you are given the structural formula of a drug molecule it is practically impossible to predict *a priori* if (i) polymorphs will be formed; (ii) how many polymorphs will be formed; (iii) whether these polymorphs will be obtained easily.
- Polymorphs can exhibit different physical and chemical properties, like color, solubility, reactivity, bioavailability, chemical and mechanical stability. This can be a boon or a bane in the pharmaceutical industry.
- Polymorphism is of major concern to industry because thermodynamic factors can change under typically accessed experimental conditions. Regulatory and legal concerns pose additional challenges.
- Competing intermolecular interactions result in dominance of kinetic factors during crystallization. A trade-off between kinetics and thermodynamics can lead to polymorphism.
- Thermodynamics and kinetics help one to assess the stability and shelf life of polymorphs.
- Polymorphs can be categorized as monotropes and enantiotropes on the basis of their thermodynamic behavior. Monotropes cannot interconvert at temperatures below the melting point. Enantiotropes can interconvert under suitable conditions.
- X-ray diffraction is the most important structural tool for the unambiguous identification of polymorphs in pure samples and mixtures. Thermal microscopy and DSC are auxiliary tools that establish thermodynamic relationships between polymorphs.

5.10   Further Reading

Books

P. Groth, *Chemische Krystallographie* (5 volumes), 1918.

L. Deffet, *Répertoire des Composés Organiques Polymorphes*, 1942.

S. R. Byrn, *Solid-state Chemistry of Drugs*, 1982.

H. G. Brittain (ed), *Polymorphism in Pharmaceutical Solids*, 1999.

J. Bernstein, *Polymorphism in Molecular Crystals*, 2002.

A. Zakrzewski and M. Zakrzewski (eds), *Solid State Characterization of Pharmaceuticals*, 2006.

R. Hilfiker (ed), *Polymorphism in the Pharmaceutical Industry*, 2006.

Papers

J. D. Dunitz, Phase transitions in molecular crystals from a chemical viewpoint, *Pure Appl. Chem.*, 63, 177–185, 1991.

T. L. Threlfall, Analysis of organic polymorphs — a review, *Analyst*, 120, 2435–2460, 1995.

J. D. Dunitz, and J. Bernstein, Disappearing polymorphs, *Acc. Chem. Res.*, 28, 193–200, 1995.

G. R. Desiraju, Crystal gazing: Structure prediction and polymorphism, *Science*, 278, 404–405, 1997.

D. Braga and F. Grepioni, Organometallic polymorphism and phase transitions, *Chem. Soc. Rev.*, 29, 229–238, 2000.

D. Giron, Investigations of polymorphism and pseudopolymorphism in pharmaceuticals by combined thermoanalytical techniques, *J. Therm. Anal. Calorim.*, 64, 37–60, 2001.

J. W. Steed, Should solid-state molecular packing have to obey the rules of crystallographic symmetry? *CrystEngComm*, 5, 169–179, 2003.

C. P. Price, A. L. Grzesiak and A. J. Matzger, Crystalline polymorph selection and discovery with polymer heteronuclei, *J. Am. Chem. Soc.*, 127, 5512–5517, 2005.

S. Aitipamula and A. Nangia, Concomitant polymorphs of 2,2′,6,6′-tetramethyl-4,4′-terphenyldiol: the β-quinol network reproduced in a metastable polymorph, *Chem. Commun.*, 3159–3161, 2005.

C. M. Reddy, S. Basavoju and G. R. Desiraju, Sorting of polymorphs based on mechanical properties. Trimorphs of 6-chloro-2,4-dinitroaniline, *Chem. Commun.*, 2439–2441, 2005.

J. Thun, L. Seyfarth, J. Senker, R. E. Dinnebier and J. Breu, Polymorphism in benzamide: Solving a 175-year-old riddle, *Angew. Chem. Int. Ed.*, 46, 6729–6731, 2007.

A. D. Bond, R. Boese and G. R. Desiraju, On the polymorphism of aspirin: Crystalline aspirin as intergrowths of two "polymorphic" domains", *Angew. Chem. Int. Ed.*, 46, 618–622, 2007.

P. M. Bhatt and G. R. Desiraju, Tautomeric polymorphism in omeprazole, *Chem. Comm.*, 2057–2059, 2007.

A. Nangia, Conformational polymorphism in organic crystals, *Acc. Chem. Res.*, 41, 595–604, 2008.

S. Crawford, M. T. Kirchner, D. Bläser, R. Boese, W. I. F. David, A. Dawson, A. Gehrke, R. M. Ibberson, W. G. Marshall, S. Parsons and O. Yamamuro, Isotopic polymorphism in pyridine, *Angew. Chem. Int. Ed.*, 48, 755–757, 2009.

D. Mangin, F. Puel and S. Veesler, Polymorphism in processes of crystallization in solution: A practical review, *Org. Proc. Res. Dev.*, 13, 1241–1253, 2009.

L. Yu, Polymorphism in molecular solids: An extraordinary system of red, orange, and yellow crystals, *Acc. Chem. Res.*, 43, 1257–1266, 2010.

A. Mukherjee and G. R. Desiraju, Synthon polymorphism and pseudopolymorphism in co-crystals. The 4,4′-bipyridine–4-hydroxybenzoic acid structural landscape, *Chem. Commun.*, 47, 4090–4092, 2011.

5.11   **Problems**

1. Write the possible tautomeric structures for 2-thiobarbituric acid. How many of these isomers are found in different polymorphic modifications of the compound? You need to go to the primary literature to obtain this answer.

2. Benzamide was thought to be monomorphic for more than 150 years despite early empirical observations by Wöhler and Liebig. Recently two other polymorphs have been characterized with X-ray crystallography. What are the structural differences in these polymorphs?

3. Resorcinol has two polymorphs α and β which melt at 108 and 110°C. Sketch a free energy versus temperature diagram similar to the one described in Section 5.3.1 for this system. Classify these polymorphs as monotropes or enantiotropes with justification.

4. The diphenol given below crystallizes from EtOAc to give two concomitant polymorphs. Form A (m.p. 257.62°C) has hydrogen bonded cyclic hexamers as in β-hydroquinone and Form B (m.p. 257.74°C) has infinite hydrogen bonded chains as in γ-hydroquinone. Form A is obtained faster than Form B and it is metastable with respect to B.

(i) Which is the kinetic form? (ii) Sketch the hydrogen bond patterns in the two forms; (iii) Provide a rationalization as to why Form A appears more quickly than Form B? (iv) Is the formation of these dimorphs governed by Ostwald's Rule of Stages?

5. Using the CSD, find out how many of the following drugs are polymorphic: atorvastatin, carbamazepine, chloramphenicol, furosemide, metformin, olanzapine, rifampicin. You may have to search in the CSD for suitable salts of these drugs. For the polymorphic drugs in this list, find out how many crystal forms and solvated forms (pseudopolymorphs) are known.

6. Oxalic acid dihydrate $(COOH)_2.2H_2O$ exists in two forms α and β. The space groups are $P2_1/n$ and $P2_1/c$. Are these forms polymorphs or pseudopolymorphs? Do you expect them to show different IR and Raman spectra? Why? Will the PXRD patterns be the same? When the β-form is cooled to 273 K it transforms to the α-form. But the α-form cannot be converted to the β-form by warming even up to 300 K. Which is the more stable form?

7. From the internet, find out more information on the legal issues concerning the polymorphs of the drugs sertraline, paroxetine and norfloxacine.

8. Read the section in this chapter on omeprazole. Verify that the tautomeric structures of this molecule are effectively the 5-methoxy and 6-methoxy isomers of the compound.

9. The following compound forms a linear one-dimensional coordination polymer. These 1D strands are arranged in a parallel manner to generate a sheet structure. Given the fact that Au(I) has a tendency to form Au⋯Au interactions, suggest possible ways in which the sheets are arranged so as to give different polymorphs.

-----Au—C≡C—⟨pyridine⟩—N—Au—C≡C—⟨pyridine⟩—N----

True polymorphism in coordination polymers is not so common. The above example is one such case. More common is what is referred to as *supramolecular isomerism* and you will learn more about this in Chapter 7.

Multi-component Crystals

6

Crystallization has always been a technique for purification. For millennia, man has crystallized salt from sea water and sugar from molasses. In the laboratory, we use crystallization to purify solid substances. Why then do some crystals contain more than one chemical constituent? All stable crystals represent free energy minima in the energy surface of some landscape. So, we may say that multi-component molecular crystals are obtained when the free energy in a system $[A_mB_n]$ is lower than what is obtained for either pure A or pure B. The energy of A_mB_n can be minimized either because of enthalpic or entropic reasons. In practice, both these conditions are realized. The types of multi-component crystals that we get in each case are quite different. Enthalpy driven multi-component crystals are characterized by distinctive intermolecular interactions. It will be possible to identify some interaction or interactions A⋯B that are more favorable than interactions of the type A⋯A or B⋯B. These crystals generally have a fixed and definite composition and the A:B stoichiometry is made up of small integers (1:1, 1:2, 2:3). Entropy driven multi-component crystals are formed when one co-crystallizes a mixture of two molecules that are of the same size, shape and chemical nature. These multi-component crystals are of the solid solution type. The crystals are of variable stoichiometry and can span entire compositional ranges. For example in A_mB_n, m and n can go all the way from 0 to 1.

6.1 General Classification and Nomenclature

It is practically impossible to have a single specific system of classification and nomenclature for all multi-component crystals. At a fundamental level, there is no difference between a single-component

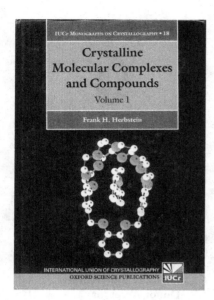

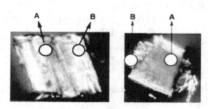

Epitaxial growth of mixed crystals

Hydrogen bonding between an organic dication and octahedral iron (A) and cobalt (B) cyanides gives two identical crystal structures. Epitaxial growth of B on a seed crystal of A or vice versa leads to the formation of a crystalline molecular alloy.

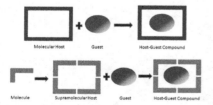

Host-guest compounds from molecular and supramolecular hosts.

crystal and a multi-component crystal. By fundamental level, we mean that the laws of crystal packing that govern the formation of single- and multi-component crystals are the same. In other words, the intermolecular interactions, and the ways in which these interactions combine with each other in the stable crystal packing, are the same. Rules for crystal design will accordingly be very similar and, in the end, even the uses to which these two types of molecular crystals are put are similar. Crystals are crystals and the tendency to distinguish multi-component crystals from single-component crystals arises largely from the fact that the concept of the molecule is so important in chemistry (Chapter 1). Crystals that contain more than one type of molecule are therefore taken up for study as a separate class of crystals, and this is what we have done in this textbook.

This having been said, there are still some differences between single- and multi-component crystals. The most important of these arises with respect to crystal growth. Since two distinct chemical compounds are involved in crystallization, one has to consider relative solubilities, vapor pressures, and their intrinsic chemical nature. A study of the phase diagram of the system is very desirable because multi-component crystals would be obtained only within certain compositional ranges.

Schemes of classification of multi-component crystals have varied from time to time and have depended to some extent on the needs and fashions of the time. Terms such as *host-guest compound* have survived over decades. Other popular terms are *clathrate, intercalate* and *inclusion compound*. These terms refer to multi-component crystals in which molecules of one compound are surrounded by molecules of another. In a clathrate, the host completely encapsulates the guest. In an intercalate the guest is sandwiched by the host in layers. In some inclusion compounds, the guest is located in one-dimensional channels in the host structure. In some types of host-guest compounds, the guest molecules can be removed to leave the undisturbed host framework (*apohost*), which can then take up another guest. In other host-guest compounds, any attempt to remove the guest results in a collapse of the host structure. These latter compounds actively require the presence of the guest during crystallization. This is referred to as guest induced host assembly (template effect). Host-guest compounds can have fixed or variable stoichiometries of the constituents.

Substitutional solid solutions are also multi-component crystals but they are mostly of variable composition. Kitaigorodskii referred to crystals of all multi-component compounds as *mixed crystals*. This term is generally not in use today. The term *molecular complex* arose in the 1960s and originally referred to cases where there is evidence for molecular association in solution. This term was especially popular for *donor-acceptor* (or charge transfer) complexes between electron rich and electron poor molecules. An even earlier usage is the term *molecular compound*. This term was applied to cases like chlorine hydrate, Dianin's compound and Hofmann's benzene compound. This term is not used today, but this terminology is quite clear and is not ambiguous.

More recent is the use of the word *co-crystal* or *cocrystal*. This term has been in use for around 20 years and there is still no complete consensus on what exactly it means. Many crystal engineers would use it to describe a multi-component crystal in which each of the components is a solid at room temperature and in which the components are associated with hydrogen bonding. The term is very popular today. Others use the term also for multi-component crystals in which the constituents are associated with interactions other than hydrogen bonding (say $\pi\cdots\pi$ interactions as in naphthalene–picric acid). Still others might use it for multi-component crystals in which one of the constituents is not a solid at room temperature. The most important subset here consists of solvates and hydrates. One of the questions posed by researchers today is whether or not a solvate is a co-crystal.

The student is well advised not to expect hard and fast definitions for various kinds of multi-component crystals. There are too many caveats in these definitions and it is possibly just as well to assume that these definitions will come and go, as they have been for the last 100 years or so.

6.2 Solid Solutions

The term *mixed crystal* was used by Kitaigorodskii for two- or multi-component crystals but he concentrated mainly on crystals built up with close-packing principles rather than with specific chemical interactions. The basis for the formation of a mixed crystal is

Dianin's compound

Some definitions of co-crystals

Only compounds constructed from discrete neutral molecular species will be considered as co-crystals. Consequently, all solids containing ions, including complex transition metal ions, are excluded.

Only co-crystals made from reactants that are solids at ambient conditions will be included. The definition can be extended to accommodate solvent (liquid) or gas as the second component.

A co-crystal is a structurally homogeneous crystalline material that contains two or more neutral molecules that are present in definite stoichiometric amounts.

Many experiments on solid solution and mixed crystal formation were conducted by Kitaigorodskii on the system diphenylmercury/tolan.

Charles J. Pedersen (1904–1989).

simple. Different atoms, ions or molecules form multi-component crystals if they are of the same size and shape because there is gain in entropy in forming such crystals. Metal alloys (brass, bronze, bell metal) have been known for millennia while intermetallic compounds have been systematically studied for nearly a century. Admixture of components is easier for inorganic crystals because atoms or ions are generally spherical. Such examples are less common among organic solids because the arbitrary sizes and shapes of organic molecules are mismatched for forming homogeneous solid solutions. Kitaigorodskii was the first to recognize that similarity of molecular structures plays a crucial role in the formation of organic solid solutions. This has been mentioned in Chapter 2. He studied the following: anthracene/acridine; diphenylmercury/tolan; 2-chloro/2-bromonaphthalene and 4,4'-bipyridyl/biphenyl. When size and shape differences between molecules increase, solid solution formation becomes restricted to narrower compositional ranges.

Other examples include mixed crystals formed by substituted benzenes with fused polynuclear aromatics like naphthalene, anthracene and pyrene. Fullerenes also form solid solutions, for example C_{60} with C_{70}. Among molecules with long n-alkyl chains such as n-alkanes, n-alcohols, fats and soaps, solid solutions appear more frequently. Pyrene based mixed crystals show interesting optical properties. Multi-component solid solutions that contain three or more different organic molecules are quite rare. Attempts to make ternary co-crystals often result in eutectic mixtures. An example of a three-component crystal containing urotropin, CBr_4 and tetrabromoadamantane has been given in Section 1.4.3.

6.3   Host-Guest Compounds

The host may be a single molecule which has a cavity, cleft or some other indentation in which can reside the guest, or it can be a supramolecular network which contains room for the guest species. The discovery of crown compounds by Pedersen initiated work in the design of molecular hosts for encapsulating guest molecules. Design of these structures is possible if one is able to understand the interactions between host and guest species. In the solid state, the host molecules interact with one another, but such contact has little to do with the host-guest interactions.

The design of a supramolecular host requires a delicate balancing of intermolecular interactions. The host network consists of several molecules that are assembled with weak intermolecular interactions such that they define a cavity, interlayer space or hollow. In a typical design strategy, one needs to take into account specific host···guest interactions and then optimize them to obtain a host-guest complex; the crystal structure will depend on the characteristics of both host and guest. Let us consider two extreme situations. In the first case, the host structure is inherently stable and complexation is enhanced with certain guests — this is similar to the behavior of zeolites. In the other second situation, the host framework is stable only in the presence of guest molecules. This is called guest induced host assembly and is reminiscent of clathrates.

As always, crystal engineering strategies for host–guest complexation depend on a proper understanding of intermolecular interactions, synthetic accessibility and a certain amount of chance. We can identify three structural attributes of a good host-guest system: (i) the host molecules should form a stable framework independent of the nature of the guest; (ii) a given host should accommodate a wide variety of guest species and; (iii) other relevant properties of the host should be capable of being tailored without any change in the complexing ability of the host.

6.3.1 *Design of Hosts*

The general tendency of organic molecules is to close-pack in crystals. Hydrocarbon crystals, in which van der Waals interactions are dominant, are close-packed. Heteroatom organic crystals, in which various kinds of directional interactions are also present, tend towards close-packing, but within the limitations imposed by directional forces. However, due to combinations of certain factors, some molecules are unable to close-pack in the usual sense. Here, a second molecule, the guest, may have to be incorporated so that the packing efficiency improves. Such factors are present when conflicting or irreconcilable demands are placed on the packing by the molecular symmetry and the directionality of intermolecular

Clathrates and organic zeolites

In organic zeolites, inclusion or sorption of guest molecules occurs without structural alteration of the host framework. Clathrates are host-guest compounds where structural deformation can occur. Zeolitic behavior is observed for host lattices which, once formed, keep their integrity whether empty or filled by guest molecules. The resulting inclusion solid is a compound of variable composition. Clathrates are characterized by host lattices which are unstable when empty. They can only exist while supported by included guest molecules. The resulting inclusion compound has an almost constant composition.

Designing Noncentrosymmetric Crystals

Functional materials are molecular systems that perform specific functions at the technological level. Examples include electronics, photonics, optoelectronics, and storage. The properties of functional molecular crystals reflect their crystal structures. Energy transfer, charge or electron transfer and proton transfer are three properties that can be fine-tuned. Although the magnitude of the properties is related to the actual molecules in the crystal, the presence or absence of the property is often controlled by the symmetry of the crystal.

Absence of an inversion centre is an important requirement for properties like pyroelectricity, piezoelectricity, tribolumi-nescence and second harmonic generation. For instance, both pyroelec-tricity and piezoelectricity require that the space group be noncentrosym-metric. A crystal is *pyroelectric* when the primitive cell possesses a dipole moment. When such a crystal is heated or cooled, there is a change in the separation of centers of positive and negative charge thereby inducing spontaneous polarization. Examples of molecular crystals are *p*-bromobenzoic anhydride and *N*-iodosuccinimide. Triglycerine sulfate is a well-known example of an organic salt that is pyroelectric. *Piezoelectricity*, which is the separation of positive and negative charge on expansion or compression of a crystal is also of technological significance. A *ferroelectric* crystal is one where a change in the direction of spontaneous polarization occurs by application of an electric field. *Tribolumines-cence* is the emission of light by a crystal on application of mechanical force. Triboluminescent crystals must have a dipole moment. Some examples are sucrose, tartaric acid, anthralinic acid and acenaphthene.

There are 21 noncentrosymmetric point groups. However, pyroelectricity is possible only in 10 of them (1, 2, 3, 4, 6, *m*, *mm*2, 3*m*, 4*mm*, and 6*mm*) as it requires alignment of all dipoles in the crystal. A ferroelectric crystal is not thus restricted because the alignment of dipoles can be altered through the influence of electric field.

Nonlinear Optical Properties

Materials with nonlinear optical (NLO) effects are useful in devices that process information efficiently. An important nonlinear effect is the *second harmonic generation* (SHG) which is used in frequency doubling of infrared and other electromagnetic radiation. Efficient chromophores (organic or metal-organic) are required to be incorporated into stable solid state assemblies to form materials exhibiting large NLO responses. Traditionally, such chromophores are noncentrosymmetric dipolar molecules containing donor and acceptor groups connected through a π backbone. Noncovalent interactions can be employed to dictate the packing of the organic chromophores in a noncentrosymmetric fashion so as to optimize the bulk properties. Organic NLO chromophores have fast response time and a high molecular polarizability (β) that can be further modulated with donor-acceptor groups on the π-conjugated framework. However, the challenge is to design crystals with a large quadratic susceptibility (χ^2). This is achieved when a molecule with a large hyperpolarizability (β) packs optimally in a noncentrosymmetric crystal.

Multi-component crystals engineered through charge transfer or hydrogen bonded interactions can show ferroelectricity. We illustrate with two examples: (i) Charge transfer in the mixed-stack, TTF- *p*-chloranil arises due to a neutral to ionic phase transition. This is an unusual example wherein intermolecular electron transfer triggers ferroelectricity with a rearrangement of molecular charge distribution. (ii) Acid-base interaction between 2,5-dihydroxy-1,4-benzoquin-one and 2,2′-bipyridine leads to a hydrogen bonded crystal. All the protons are in long-range order so that the O–H and N–H$^+$ bonds are all aligned in the same direction and the alternate O–H$\cdots$N and N–H$^+\cdots$O$^-$ bonds constitute a polar chain. When all the protons are transferred simultaneously, the chain is reversed in polarity (*p*) without losing its chemical identity. Ferroelectricity arises due to the collective proton transfer process under an external electric field.

(a)

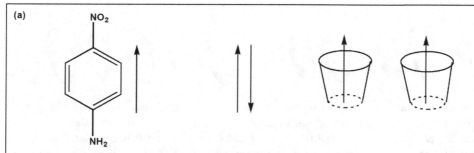

Inclusion complex of 4-nitroaniline with cyclodextrin results in as alignment of dipoles to give a noncentrosymmetric structure. The native crystal structure is centrosymmetric.

(b)

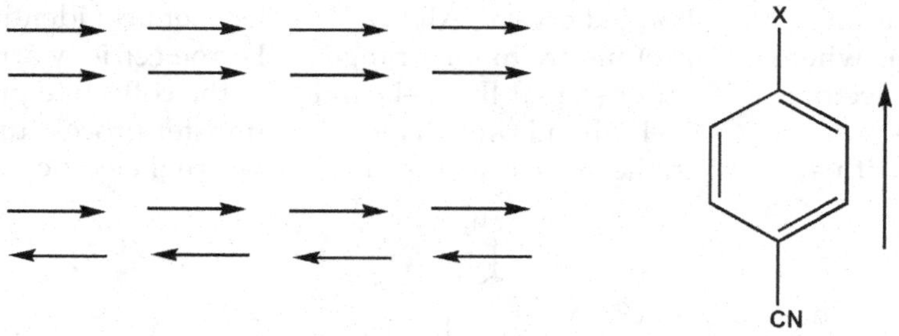

X = Cl, Br, I

Substituted benzonitriles can take, in principle, polymorphic structures. Some arrangements make the crystal NLO active, while others result in a centrosymmetric structure.

(c)

A co-crystal between two polar molecules, 4-nitro,4'-hydroxybiphenyl and 4-nitrosotoluene can result in two configurations of which one is centrosymmetric and the other is noncentro-symmetric due to the arrangement of dipoles. The alignment of major diploes in a centrosymmetric manner gives a statistical chance for an overall noncentrosymmetric packing.

(d)

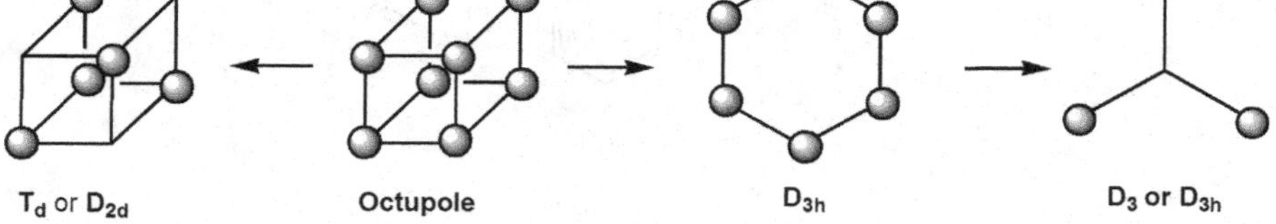

T_d or D_{2d} Octupole D_{3h} D_3 or D_{3h}

Another approach is based on octupolar molecules. Trigonal and tetrahedral octupolar molecules, by definition, have a zero dipole moment, but the crystal structure permits a macroscopic SHG.

interactions. In such cases, relatively open host networks often result. An understanding of the factors that render close-packing of a single-component crystal difficult is the key to systematic crystal engineering of host-guest compounds.

The symmetry of a molecule is almost always lowered in the crystal with the inversion centre being the only normally retained molecular symmetry element. For a low-symmetry molecule, this is hardly a problem and crystal packing is accomplished with the use of twofold screw axes and glide planes. Higher symmetry molecules generally do not adopt higher crystal symmetries because such symmetries are incompatible with the geometries of the common types of intermolecular interactions such as the herringbone, the hydrogen bond and other weakly directional forces described in Chapter 2. Therefore many molecules take low-symmetry crystal systems and close-pack as single-component crystals. For example, benzene, coronene and hexamethylbenzene are all sixfold symmetrical molecules but they adopt orthorhombic, monoclinic and triclinic crystal packing respectively.

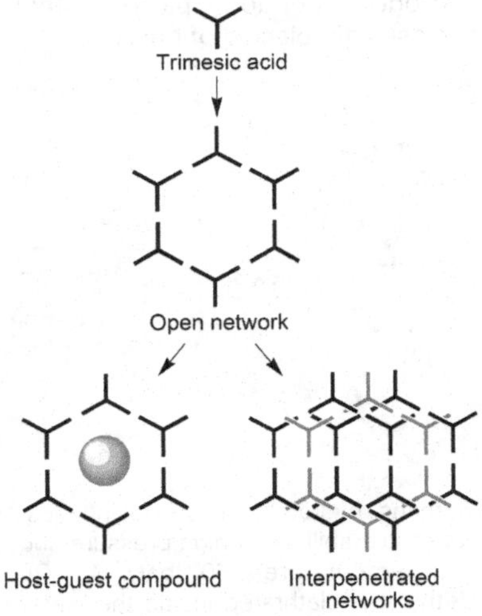

The tendency towards conflicting factors in crystal packing is pronounced when groups capable of hydrogen bonding are symmetrically disposed in a symmetrical molecular framework. The rigid scaffolding of hydrogen bond is also compelled to be symmetrical because the directionality of these hydrogen bonds cannot be compromised. Gross violations of the close-packing principle may result, leading to the formation of an open host network. These open networks can either interpenetrate (Section 7.6) or may include guest molecules. Good examples are provided by the clathrates of hydroquinone (1,4-dihydroxybenzene) and trimesic acid (benzene-1,3,5-tricarboxylic acid). The latter forms either host-guest complexes or interpenetrated structures, depending on the conditions employed during crystallization.

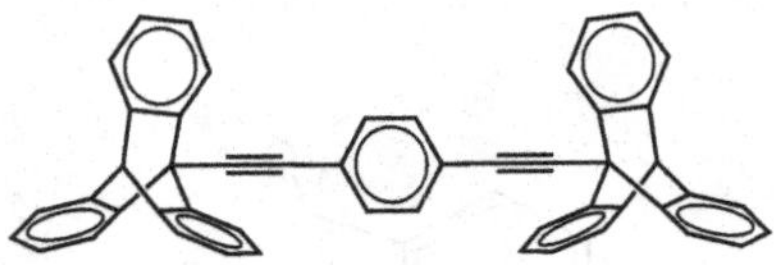

Awkwardness in molecular shape hinders close-packing. A good example is provided by the so-called wheel-and-axle molecules. These molecules contain a long linear molecular axis composed of *sp* carbon atoms (axle) bearing at both ends large rigid substituent groups containing *sp*^{3} hybridized C-atoms (wheels). The wheels prevent the close-packing of the axles and voids are created. The shapes and sizes of the voids, hence the nature of the guest molecules, can be varied by tuning the axles. Examples include many dialkynes and allenes.

Trimesic acid forms open and inter-penetrated networks.

Wheel and axle compound

Supramolecular host: Enol form of 1,3-cyclohexanedione forms a cyclic hydrogen bonded pattern that encloses a molecule of benzene.

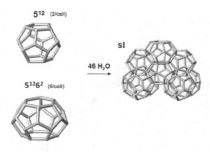

Methane hydrate is a crystalline, ice-like solid stabilized at high pressure and low temperature (60 bar, 4 °C). Methane is clathrated inside the water cages. The structure is made up of two smaller cages (5^{12} and $5^{12}6^2$).

Saccharin

(*Continued*)

6.4   Solvates and Hydrates

The inclusion of crystallizing solvents in the crystal structure of a molecular solid takes place for a variety of kinetic and thermodynamic reasons. Most non-ionic organic crystals (>80%) are unsolvated because the entropic gain in eliminating solvent molecules from the crystal nuclei into the bulk solution overrides any enthalpic gain that is obtained from solute···solvent interactions. Conversely, when interactions between solute and solvent are unusually strong and involve multi-point recognition, enthalpic considerations become important, and the solvent is retained in the crystal. Whether these solvates and hydrates can be termed host-guest complexes depends largely on the details of the individual crystal structures, that is, on whether a host network can be clearly identified. In practice, a large number of solvents (water, methanol, ethanol, ether, benzene, acetone and chloroform) are included in crystals. Polar solvents like water are almost always stabilized in the crystal structure through directional interactions. The tendency for a polar solvent to be retained in crystals often follows from its ability to form multi-point recognition hydrogen bonding synthons with the solute. This was shown with a CSD analysis several years ago. In contrast, solvents that have only a single hydrogen bonding site are not included so often in crystals. Non-polar solvents fulfill a space-filling role. The phenomenon of solvation or hydration is connected with polymorphism and is of industrial importance since the properties of many drugs and dyestuffs seem to be dependent on the properties whether or not solvent is included in the crystal.

The inclusion of water in molecular solids is a matter of fundamental and practical importance and is quite unlike the incorporation of other solvents of crystallization. Due to its small size and excellent hydrogen bonding capability, water plays a critical role in many crystal structures. In the context of host guest chemistry, water is important both as a host as well as a guest. Clathrate hydrates are excellent examples wherein water behaves like a host. However, water mostly occurs as a guest species in hydrates. A notable reason for hydration is that whenever an organic molecule contains more hydrogen bond acceptor groups than donor groups, water can perform an important role in balancing the number of two-centered hydrogen bonds. The water molecule is one of very few common molecules in which the donor atoms outnumber the acceptor atoms: the formation of a hydrate by acceptor rich molecules therefore reduces the need to form three- and four-centered (multifurcated) hydrogen bonds. Water is also unusual in that it is the only solvent that may occur in a crystal structure even though it is not specifically used in the crystallization experiment. There is enough water vapor in the atmosphere under ambient conditions so that water can always be included in a crystal if there are specific chemical reasons for so doing. For both these reasons, a very large number of hydrates are known. The CSD contains at least 5000. Water is found around ten times more frequently than methanol, the second most commonly found solvent in molecular crystals.

Other solvents are included either because they can form specific hydrogen bonds with the host compound. Examples are methanol, ethanol, acetone, ether, dioxane, DMF and pyridine. Non-polar solvents are included because they can fill voids and cavities in the crystal packing. Such solvents include benzene, CCl_4 and xylenes.

6.5 Donor-Acceptor Complexes

Unlike solid solutions, these complexes have definite stoichiometries. The physical and chemical properties of these complexes are quite different from the individual molecules from which they are made. Donor-acceptor complexation generally occurs between π-donors and π-acceptors. Some workers also refer to these complexes as co-crystals. The formation of strong complexes is associated with enhancements in aromatic ring polarization in the constituent molecules, either electron depletion or excess. For example, an increase in the nitro groups on an acceptor molecule (picric acid) or in the number of alkyl or alkoxy groups on a donor molecule (durene) will increase the complexing ability.

In the crystal engineering context, donor-acceptor complexation is most often associated with electrical conductivity and superconductivity. Planar aromatic donor and acceptor molecules stack in either mixed or segregated stacks. Design strategies are usually directed towards the latter as it is the segregated stack arrangement that is associated with anisotropic electronic conductivity. Stacking is almost always accompanied by lateral offset of planar rings in adjacent electron-rich and

(Continued)

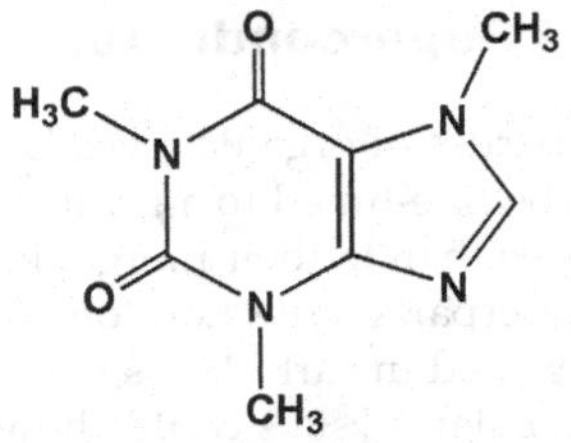

Caffeine

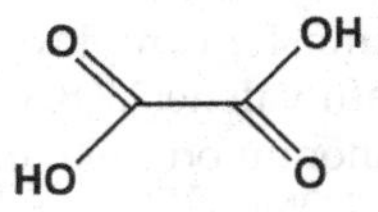

Oxalic acid

Some organic molecules that exist as hydrates

Pseudopolymorphism

In general, organic crystals do not include solvent during crystallization. This observation may be rationalized by assuming that crystallization begins with solute-solvent aggregates that contain solute-solute, solute-solvent and solvent-solvent interactions. The entropic gain in eliminating solvent molecules from these aggregates into the bulk solution, and the simultaneous enthalpic gain in forming stable solute species that contain robust supramolecular synthons provides an adequate driving force for the crystallization of solvent-free crystals. When solvent is included, one needs to consider the formation of pseudopolymorphs, that is crystalline forms of a compound that differ in the nature or stoichiometry of the included solvent. Inclusion of solvent in organic crystals depends on the nature of both solute and solvent. If solute-solvent interactions are unusually important, say because of multipoint recognition, the entropic advantage associated with solvent expulsion into the bulk may be overridden by these additional enthalpic factors resulting in retention of some solvent in the crystal. For example, 3,5-dinitrosalicylic acid shows four pseudopolymorphs with dioxane; different supramolecular synthons (shown below) are responsible for the formation of these solids.

Molecular Conductors and Superconductors

Properties of organic crystals can be fine-tuned to a greater degree than in their inorganic counterparts. Fritz London suggested in early 1950s that molecular crystals could show superconductivity. However, the first organic metal, a charge transfer complex, tetrathiafulvalene:7,7,8,8,-tetracyanoquinodimethane ($TTF^{+0.59}TCNQ^{-0.59}$) was realized only in 1970s. Organic metals are typically 4 Å short-axis donor-acceptor complexes with segregated stacks of planar aromatic π donors (D) and π acceptors (A). These D and A molecules are arranged in mixed or segregated stacks. Molecular stacks favor interaction of π-electrons resulting into a range of energy levels or bands instead of discrete energy levels associated with isolated molecules. This in turn leads to appreciable electronic conductivity along the stacks when electrons are removed or added to the energy bands. While mixed stacks favor semiconducting behavior, segregated stacks can result in metallic or superconducting behavior (TTF-TCNQ). Small changes in the molecular structures of D and A enhance lateral contacts which is desirable because these contacts increase the structural dimensionality and thereby the conducting or superconducting property. Such thinking led to the first organic superconductor, $(TMTSF)_2PF_6$ ($T_c \sim 1.2$ K). At low temperatures TTF-TCNQ undergoes a phase transition from the conducting to a semiconducting phase caused by a pairing of donor and acceptor stacks. Conductivity depends on the interplanar separation or uniform stacking which in turn affects orbital overlap and hence band structure. TTF-TCNQ forms uniform stacks and shows metallic behavior at room temperature. TTF-TCNQF$_2$, on the other hand show dimerized stacks (ring-over-ring) leading to semiconducting behavior.

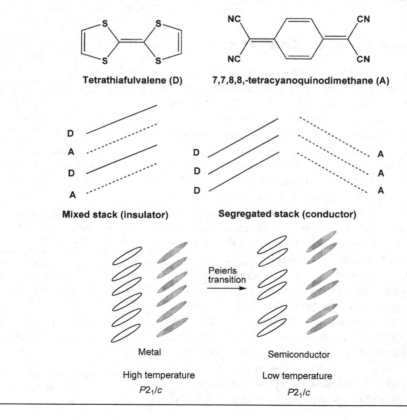

Tetrathiafulvalene (D)   **7,7,8,8,-tetracyanoquinodimethane (A)**

Mixed stack (insulator)   Segregated stack (conductor)

Peierls transition

Metal   Semiconductor

High temperature   Low temperature

$P2_1/c$ $P2_1/c$

electron-deficient molecules. This offset is important in certain types of molecular superconductors based on the donor molecule BEDT-TTF. A very important type of lateral interaction in donor-acceptor complexes is O–H···O hydrogen bonding. The stronger π-π complexes with fixed stoichiometries are usually characterised by some type of hydrogen bonding. Quinone and hydroquinone molecules alternate along a linear hydrogen bonded chain and the chains themselves are linked with C–H···O interactions (Section 2.3). Hexamethylbenzene and 1,3,5-tricyanobenzene form a 1:1 molecular complex. Here each layer contains only a single-component. The stacking of the layers one over the other is optimised with charge transfer interactions between donor and acceptor.

Donor-acceptor complexation can be utilized to design new photoreactive crystals. An example of a donor-acceptor complex engineered for a topochemical reaction is the 1:1 complex formed between 3,5-dinitro- and 3,4-methoxycinnamic acids. The structure is made up of π-stacked molecules as a result of O–H···O and C–H···O hydrogen bonding combined with donor-acceptor charge transfer interactions. The 1:1 molecular complex undergoes dimerization in the presence of UV-light to give a cyclic product. The acid formed is of the β-truxinic type (Chapter 1) because the donor-acceptor complex formation brings adjacent electron deficient and electron rich rings into close proximity. Another related example of an engineered [2 + 2] photocycloaddition reaction that uses donor-acceptor complexation as a design tool is furnished by the 1:1 complex between resorcinol and *trans*-1,2-bis(4-pyridyl)ethylene (4,4'-bpe). The product is a discrete, four-component complex, held together with four O–H···N hydrogen bonds and the stacking occurs orthogonal to the diol. This assembly gives a favorable geometry to induce photochemical [2 + 2] cycloaddition reaction. A similar strategy can be used with the donor and acceptor functionalities reversed. For example, the 1:1 complex between fumaric acid and 2,3-bis(4-methylenethiopyridyl)naphthalene forms *rctt*-1,2,3,4-cyclobutanetetracarboxylic acid under UV irradiation.

6.6 Co-crystals

In Section 3.3, we learned about the utility of the synthon approach in the design of crystal structures of

Layer structure parallel to (111) of 1,3,5-tricyanobenzene molecules in the crystal structure of its co-crystal with hexacyanobenzene. This structure is also discussed in Chapter 3.

molecular solids. In this section, we look at a particular class of multi-component solids, namely, the *co-crystal*. What distinguishes a co-crystal from other categories of multi-component solids? As mentioned earlier in this chapter, the definition of a co-crystal is not a hard and fast one, but there is a certain emphasis on the interactions that bind the dissimilar molecules that constitute a co-crystal. Accordingly, when a multi-component solid is defined as a co-crystal, there is an association to design strategies that may be used to make such a multi-component system. The incorporation of multiple components within a crystal gives chemists flexibility to deliberately modify the composition and the complexity of a solid. How do we extend the synthon concept to the design of a co-crystal? First let us look at some binary halogen atom systems like 4'-bromo-4-cyanobiphenyl/4'-bromo-4-cyanobenzene and methyl-1,4-dicyanobenzene/2,3,5, 6-tetrafluoro-1,4-diiodobenzene. In these cases, a 1:1 co-crystal is directed through $Br \cdots N \equiv C$ or $I \cdots N \equiv C$ intermolecular interactions aligning the two components in head-to-tail fashion. The synthon here is a single interaction. Other interactions such as nitro$\cdots$iodo, cyano$\cdots$nitro and halogen$\cdots$halogen can also direct structures. These interactions are of limited use for design purposes. On the contrary, multi-point synthons based on hydrogen bonds are stronger and more reliable. Hence they can function as ideal tools for crystal engineering. In the following section, we will examine how hydrogen bonds can be utilized to design co-crystals.

6.6.1   *Hydrogen Bonded Co-crystals*

Two of the most common families of hydrogen bonded compounds are carboxylic acids and primary amides. Let us consider the crystallization of a two component system containing, say a carboxylic acid and an amide. Let us first analyze what we can infer from the individual components. In Section 3.3.2, we learned that carboxylic acid containing crystal structures are based on two alternative patterns: homodimer and catemer. A primary amide, say acetamide also shows similar (competing) patterns, a homodimer and a catemer. The acid–acid and amide–amide homodimer synthons are called *homosynthons*. What happens if we combine an acid and an amide to form a co-crystal? Two possibilities arise: the co-crystal contains homosynthons of the kind discussed above or there is a new

acid-acid homosynthon

amide-amide homosynthon

acid-amide heterosynthon

synthon made up of an acid and an amide fragment. This synthon is called a *heterosynthon*. Identification of reliable interactions between multiple components through heterosynthons is the underlying concept for hydrogen bonded acid amide co-crystal synthesis.

The classification of synthons as homosynthons (containing the same set of functional groups) and heterosynthons (containing different sets of functional groups) was first suggested by Zaworotko and provides a retrosynthetic approach to analyze the structures of co-crystals. An examination of several multi-component solids containing acid and amide fragments suggests that heterosynthons occur more frequently than homosynthons. Typical examples of co-crystals containing acid–amide based heterosynthons are: succinic acid–benzamide and urea–glutaric acid. The reliability and robustness of this synthon is

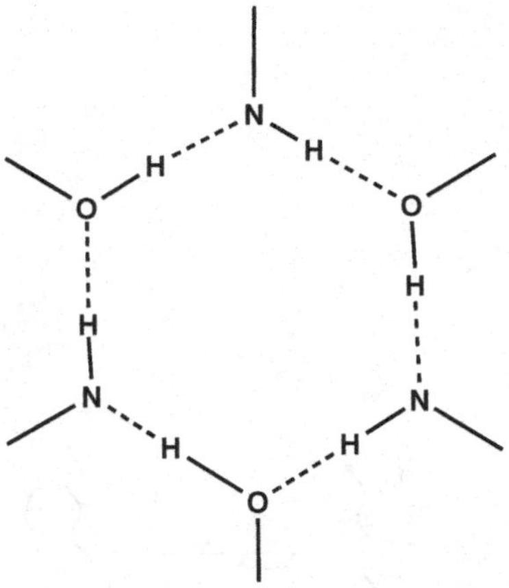

β-As N–H···O–H network

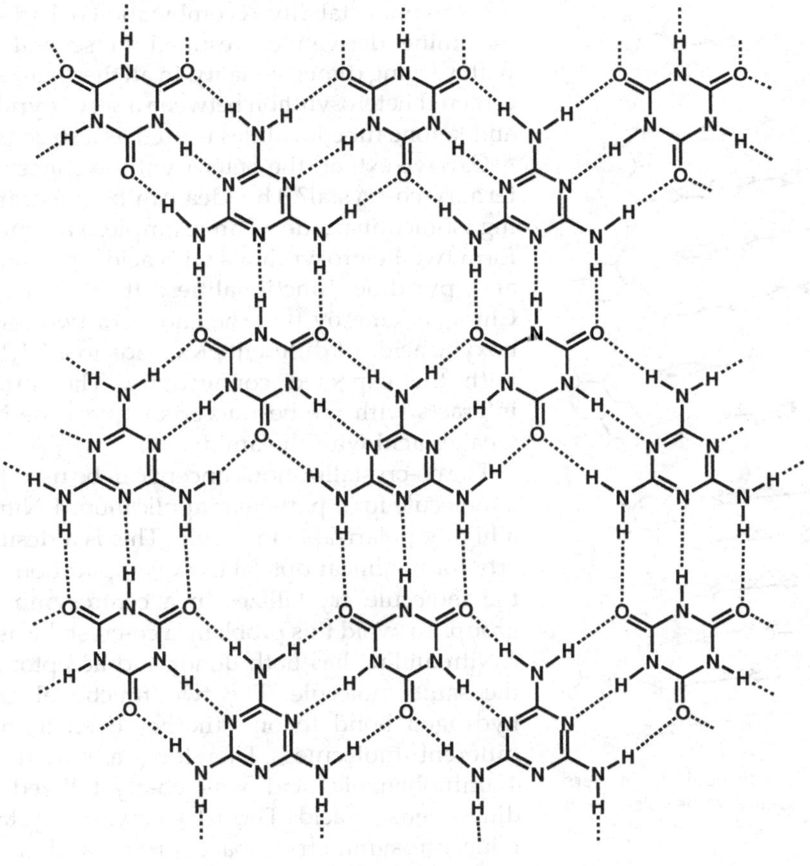

A chicken-wire pattern

exploited in the formation of a pharmaceutically viable co-crystal, carbamazepine-saccharin.

Combining alcohols and amines is another way to integrate multiple components into a crystal. An alcohol has one hydrogen bond donor and two acceptors (one for each lone pair). An amine is perfectly complementary to it with two hydrogen bond donors and one acceptor. A heterosynthon of this kind can yield a chicken-wire or a ladder structure wherein all potential hydrogen bond sites are satisfied.

The preceding examples illustrate how heterosynthons between two interacting centers can be exploited to design co-crystals. In many cases, these operate between strong donors and acceptors. Whenever we have weaker donors and acceptors, a multicentered approach becomes effective. George Whitesides and J.-M. Lehn approached co-crystal design with the notion that multi-centre hydrogen bonds would provide more stable assemblies. In addition, complementarities were seen as a key to predictability. A combination of barbituric acid/melamine derivatives resulted in several co-crystals with varying dimensionality. In all these cases, the three-centered heterosynthon between a set of pyridine, amide and ketone functionalities is used as a structure director.

Can we extend the heterosynthon concept to form a ternary co-crystal? The idea can be illustrated by taking isonicotinamide as an example. This molecule can form two heterosynthons with acids through its amide and pyridine functionalities. It was suggested by Christer Aakeröy that the choice of two different carboxylic acids of different pKa leads to a 1:1:1 co-crystal with the expected connectivity. The stronger acid interacts with the best acceptor (pyridine N) and the weaker acid with the amide.

The co-crystallization concept can be used to engineer a molecule for a particular application. 4-Nitroaniline is a highly polarizable molecule. This is a desirable property for nonlinear optical (NLO) applications. However, the molecule crystallizes in a centrosymmetric space group. To avoid this problem, a co-crystal was designed. 4-Nitroaniline has both donor and acceptor groups on the same molecule. The two functional groups will hydrogen bond to one another even if they are on different molecules. Therefore, a structural analog 4-aminobenzoic acid was co-crystallized with 3,5-dinitrobenzoic acid. The 1:1 co-crystal crystallizes in a noncentrosymmetric space group and shows large second order hyperpolarizability in the solid state.

Homologous (isoreticular) ladders formed by 3-aminophenols with bipyridines.

6.6.2 *Pharmaceutical Co-crystals*

Co-crystals are of special importance in the pharmaceutical industry. Because drugs are usually sold as solid formulations, the solid-state chemistry of active pharmaceutical ingredients (API) has always been an integral part of drug development. The main idea behind the development of a *pharmaceutical co-crystal* is simple. A two component crystal is prepared, consisting of a drug and an auxiliary compound known as a *co-former*. The co-former is selected from a list of compounds that are generally recognized as safe (GRAS). The co-crystal has some property or properties that make it more advantageous in formulation, when compared with the original drug. This favorable property may be enhanced solubility.

Co-crystals are generally more suited to drug formulation than are solvates and hydrates. This is because the number of pharmaceutically acceptable solvents is limited. Further, solvents tend to be volatile and solvates are susceptible to desolvation leading to amorphization, which can be undesirable. In contrast, most co-formers are unlikely to evaporate from solid dosage forms, making phase separation and other physical changes less likely.

Pharmaceutical co-crystals lend themselves well to patent protection, unlike polymorphs. The three criteria of patentability are novelty, non-obviousness and utility. Pharmaceutical co-crystals generally satisfy all three criteria admirably. A co-crystal usually satisfies the novelty criterion because it is a new composition of matter substance. Non-obviousness is provided by the fact that the identification of the co-former is hardly ever routine, unlike say salt formation wherein an acid is obviously required to make a salt from a base. Utility is generally the only criterion that must be established but it is often easy to demonstrate — usually it is the lack of a particular attribute (solubility, bioavailability, dissolution profile) that has led to the identification of a pharmaceutical co-crystal.

6.6.2.1 *Design of Pharmaceutical Co-crystals*

There are two main approaches to the design of pharmaceutical co-crystals. The first uses a retrosynthetic strategy and the structure of the co-former is deduced by identifying complementary hydrogen bonding sites

Co-crystallization of 4-aminobenzoic acid with 3,5-dinitrobenzoic acid.

Pharmaceutical formulations are generally physical mixtures of an active pharmaceutical ingredient (API) and an inactive ingredient (excipient) used as a carrier. The nature of the physical form and its formulation exerts a profound effect on the bioavailability parameters, especially for water insoluble compounds.

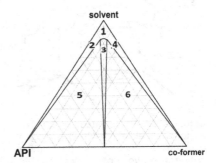

Ternary phase diagram for the system API, co-former and solvent. Region **1** represents the undersaturated region bounded by the solubility curves of API and co-former. Regions **3**, **5** and **6** represent the stable solid phases of co-crystal, API and co-former respectively.

The concept of non-obviousness is to a large extent, a legal matter. As per U.S. patent law, a patent may not be obtained "*if the differences between the subject matter sought to be patented and the prior art are such that the subject matter as a whole would have been obvious at the time the invention was made to a person having ordinary skill in the art to which said subject matter pertains.*" Unlike salt forms, the formation of co-crystals is a considered to be a non-obvious matter and is thus eligible for patent protection. This because the design of co-crystals involves rational thinking, reasoning ability and a knowledge of the intermolecular interactions involved in the process. There is also no guarantee that co-crystallization of a drug and a co-former will result in the formation of a co-crystal.

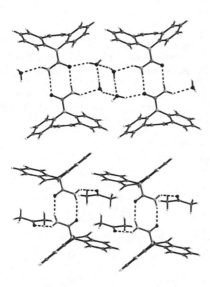

Carbamazepine solvates: water or acetone is held by N–H···O hydrogen bonding and occupies the space between two pairs of amide homosynthons.

in the API and co-former. The oral administration of carbamazepine, an anti-epileptic drug, faces many challenges: low aqueous solubility, low bioavailability and therefore high dose requirements. The molecule contains an amide group and the crystal structure of the native drug contains an amide dimer homosynthon. Two retrosynthetic possibilities suggest themselves. The first involves the making of an amide–acid heterosynthon in lieu of the amide homosynthon in the native structure. So, carbamazepine forms co-crystals with carboxylic acids such as acetic acid, aspirin, benzoic acid, formic acid, malonic acid and succinic acid. The second synthetic route retains the amide homosynthon but utilizes other hydrogen bond donor and acceptor sites in the drug molecule. Strong acceptors like carbonyl compounds or sulfoxides and donors such as alcohols are appropriate as co-formers. This strategy enables carbamazepine to form co-crystals with saccharin, 1,4-benzoquinone and 4-aminobenzoic acid. All these co-crystals show superior dissolution properties, suspension stability and pharmacokinetics in comparison with the parent compound.

The second approach to the preparation of pharmaceutical co-crystals is based on high throughput crystallization. Solution based crystallization is an option, but suffers from the long times involved and poor efficiency. Mechanochemical methods provide an alternative path (Section 4.1.5.2). While traditional solution-based screening requires a great variety of solvent systems and crystallization conditions, thermal and mechanical methods provide enhanced experimental efficiency.

A combination of the two approaches may prove advantageous.

6.6.2.2 *Properties of Pharmaceutical Co-crystals*

Function and performance determine success or failure of pharmaceutical formulation. Co-crystals of glucuronic acid with caffeine, norfloxacin with isonicotinamide, and itraconazole with succinic acid, L-malic acid and L-tartaric acid exhibit higher solubilities than the pure drugs. Co-crystallization can also be utilized as a strategy for decreasing the solubility. 1:1 or 1:2 caffeine-gentisic acid co-crystals are less soluble than pure caffeine. These complexes are used to formulate caffeine in dosage forms such as chewable tablets that are intended to linger in the mouth. Such dosage forms release caffeine slowly and have an improved taste factor over ones containing pure caffeine.

The stability of a solid drug substance in the presence of atmospheric moisture is important for processing, formulation, packaging, and storage. For example, caffeine is susceptible to hydration with change in humidity. The co-crystal of caffeine with oxalic acid provides stability to the drug molecule towards humidity for several weeks. Mechanical properties of molecular crystals are a function of intermolecular interactions and crystal packing. Paracetamol, a widely used analgesic drug is dimorphic. The metastable Form II consists of hydrogen bonded layers; this form has superior compaction properties. The thermodynamically stable Form I, which is composed of corrugated hydrogen-bonded layers of molecules, invariably crumbles upon compaction. Form II is obviously preferred for tableting but it is less stable than Form I. Co-crystals of paracetamol with oxalic acid and theophylline are stable but they form layered structures that resemble Form II and they are compacted easily.

Co-crystal formation is also possible between two drug molecules. An example is that of the co-crystal that is formed between the two anti-HIV drugs, lamivudine and zidovudine, marketed under the trade name Combivir. The co-crystal may be more effective than the individual drugs in therapeutics.

6.6.2.3 *Co-crystals and Salts*

Pharmaceutical co-crystals are generally formed between an acid and a base because many drugs are basic in nature. The operative interaction in co-crystal formation is a hydrogen bond, say of the type O–H···N. Any hydrogen bond is an incipient proton transfer. If the acidity and basicity of the two components is sufficiently marked, proton transfer can occur across the hydrogen bond and a charge assisted hydrogen bond of the type O^-···$H–N^+$ hydrogen bond results.

A simple guideline is available for choosing acids and bases for salt and co-crystal formation. Let us consider the salt formation in a three component system containing an acid (HA), a base (B) and a solvent. A salt is formed by transfer of a proton from the acid to the base. Proton transfer depends mainly on the pKa values of the two components. Salt formation generally requires a difference of at least 2.7 pKa units between the base and the acid. For example, succinic acid with a pKa of 4.2 forms a co-crystal with urea which has a pKa of 0.1. On the other hand, it forms a salt with L-lysine (pKa 9.5).

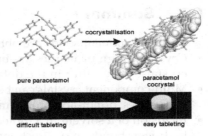

Co-crystallization of paracetamol leads to better compaction properties.

Lamivudine   Zidovudine

Co-crystal between succinic acid and urea.

6.7 Summary

- Multi-component solids are obtained due to enthalpic or entropic reasons.
- Crystal design principles are broadly the same for single- or multi-component target crystals. They differ in the details for the two varieties of solids.
- Multi-component crystals can be host-guest compounds, clathrates, inclusion compounds, intercalates, donor-acceptor complexes or co-crystals.
- Inclusion of a solvent in a molecular crystal generally owes to the formation of specific hydrogen bonds and directional interactions, or because of close-packing in cavities and voids. Solvation can be often predicted.
- Donor-acceptor complexes are formed between π-donors and π-acceptors.
- Hydrogen bonded co-crystals between acids, amides and alcohols are very common. The concept of homosynthons and heterosynthons provide design strategies to form new co-crystals.
- Other nonbonding interactions like nitro⋯iodo and cyano⋯nitro can also be used direct co-crystal formation.
- Design of a host-guest compound is based on three structural attributes: a stable host, its ability to accommodate a variety of guest and tailorability of the host without altering its basic structure.
- Co-crystal formation can direct functionality to a solid, for example in organic conductors and NLO materials.
- Co-crystal formation can be exploited to carry out solid state reactions like photodimerization.
- Pharmaceutical co-crystals are generally based on acid-base interactions.
- Pharmaceutical co-crystals can extend the patentability of a drug because they often satisfy the triple criteria of novelty, non-obviousness and utility.
- A guideline suggests that if ΔpKa > 2.7 for an acid and a base, co-crystallization leads to salt formation rather than to a co-crystal.

6.8 Further Reading

Books

R. Foster, *Organic Charge-Transfer Complexes*, 1969.
A. I. Kitaigorodskii, *Mixed Crystals*, 1984.
F. H. Herbstein, *Crystalline Molecular Complexes and Compounds* (2 volumes), 2005.

Papers

A. Nangia and G. R. Desiraju, Pseudopolymorphism: Occurrences of hydrogen bonding organic solvents in molecular crystals, *Chem. Commun.*, 605–606, 1999.
L. R. MacGillivray, J. L. Reid and J. A. Ripmeester, Supramolecular control of reactivity in the solid state using linear molecular templates, *J. Am. Chem. Soc.*, 122, 7817–7818, 2000.
C. B. Aakeröy, A. M. Beatty and B. A. Helfrich, "Total synthesis" supramolecular style: Design and hydrogen-bond-directed assembly of ternary supermolecules, *Angew. Chem. Int. Ed. Engl.*, 40, 3240–3242, 2001.
Ö. Almarsson and M. J. Zaworotko, Crystal engineering of the composition of pharmaceutical phases. Do pharmaceutical co-crystals represent a new path to improved medicines? *Chem. Commun.*,1889–1896, 2004.

B. R. Bhogala, S. Basavoju and A. Nangia, Tape and layer structures in cocrystals of some di- and tri-carboxylic acids with 4,4′-bipyridines and isonicotinamide. From binary to ternary cocrystals, *CrystEngComm*, 7, 551–562, 2005.

D. P. McNamara, S. L. Childs, J. Giordano, A. Iarriccio, J. Cassidy, M. S. Shet, R. Mannion, E. O'Donnell and A. Park, Use of a glutaric acid cocrystal to improve oral bioavailability of a low solubility API, *Pharm. Res.*, 23, 1888–1897, 2006.

L. S. Reddy, N. J. Babu and A. Nangia, Carboxamide-pyridine *N*-oxide heterosynthon for crystal engineering and pharmaceutical cocrystals, *Chem. Commun.*, 1369–1371, 2006.

J. A. Bis, O. L. McLaughlin, P. Vishweshwar and M. J. Zaworotko, Supramolecular heterocatemers and their role in cocrystal design, *Cryst. Growth Des.*, 6, 2648–2650, 2006.

S. L. Childs, G. P. Stahly and A. Park, The salt-cocrystal continuum: The influence of crystal structure on ionization state, *Mol. Pharm.*, 4, 323–338, 2007.

A. V. Trask, An overview of pharmaceutical cocrystals as intellectual property, *Mol. Pharm.*, 4, 301–309, 2007.

C. B. Aakeröy, S. Forbes and J. Desper, Using cocrystals to systematically modulate aqueous solubility and melting behavior of an anticancer drug, *J. Am. Chem. Soc.*, 131, 17048–17049, 2009.

L. Rajput and K. Biradha, Design of cocrystals via new and robust supramolecular synthon between carboxylic acid and secondary amide: Honeycomb network with jailed aromatics, *Cryst. Growth Des.*, 9, 40–42, 2009.

A. Ainouz, J.-R. Authelin, P. Billot and H. Lieberman, Modeling and prediction of cocrystal phase diagrams, *Int. J. Pharm.*, 374, 82–89, 2009.

6.9 Problems

1. 1,4-dichlorobenzene and 1,4-dibromobenzene form solid solutions in all compositions and these crystals are isomorphous with crystals of 1-chloro-4-bromobenzene. However, none of these compounds form solid solutions with *p*-xylene, although the volumes of the chloro and methyl groups are comparable. The crystal structure of *p*-xylene is quite distinct from that of 1,4-dichlorobenzene. On the other hand, the crystal structure of 4-chloronitrobenzene, in which the chloro and nitro groups are disordered, is rather similar to that of 1,4-dichlorobenzene. Suggest reasons for these observations.

2. Durene (1,2,4,5-tetramethylbenzene) forms a solid donor-acceptor (charge transfer) complex with 1,3,5-trinitrobenzene but not with nitrobenzene. Give a reason for this.

3. Are the following likely to form co-crystals, solid solutions or solvates? Or will they crystallize separately? Rationalize your answer: (i) CCl_4 and CBr_4; (ii) $C(p\text{-}C_6H_4Cl)_4$ and $C(p\text{-}C_6H_4Br)_4$; (iii) CCl_4 and $C(p\text{-}C_6H_4Cl)_4$; (iv) Suberic and sebacic acids, $HO_2C(CH_2)_6CO_2H$ and $HO_2C(CH_2)_8CO_2H$; (v) C_6H_6 and C_6F_6; (vi) Compounds **A** and **B**.

A   B

4. Consider the eight two-component crystals below. All of them have been isolated and the crystal structures have been determined. Estimate which ones were obtained as co-crystals and which ones as salts.

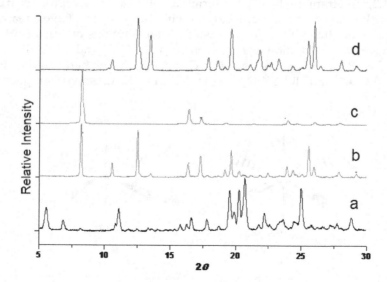

5. You are given four PXRD traces of (i) benzoic acid; (ii) 4,4'-bipyridine; (iii) physical mixture of the two compounds; (iv) 1:1 co-crystal of the two compounds. Select from these four traces, the two which represent the pure compounds. Select the trace that corresponds to the physical mixture. Select the one that corresponds to the co-crystal.

6. 5-Nitrosalicylic acid crystallizes as an anhydrate from water, but 3,5-dinitrosalicylic acid readily forms a hydrate. In fact, it has not been obtained till now in an anhydrous form. Suggest a reason for these observations.

7. Suggest co-formers for the following drugs: alprazolam, aspirin, fluconazole, isoniazid, lamivudine, temozolomide.

Coordination Polymers

7

Coordination polymers are a varied group of crystalline coordination compounds that lend themselves easily to structural design principles. They are molecular solids in the sense that Kitaigorodskii intended: within the crystallographic unit cell, one can identify groups of atoms, the distances between which are smaller than the distances between an atom within this group and an atom in another group. Coordination polymers also have many interesting and useful properties, which the chemist can fine tune. All these reasons justify the study of coordination polymers within the subject of crystal engineering.

The dark blue paint pigment Prussian Blue was first made in Berlin around 1700. Since then, this blue pigment has been widely used in a number of paintings all over the world. There are different forms of Prussian blue; the soluble form has the formula $KFe^{III}[Fe^{II}(CN)_6]$ while the insoluble form is $Fe^{III}_4[Fe^{II}(CN)_6]_3 \cdot xH_2O$. The crystal structure of the insoluble form was determined only in 1977, when it was realized that the pigment has a three-dimensional polymeric structure. Today, it would be considered to be a coordination polymer.

7.1 What are Coordination Polymers?

Coordination polymers constitute a special case of the very large and diverse group of substances that we call coordination compounds or coordination complexes. In 1893, Werner proposed that a coordination complex consists of a central metal ion (or ions) bonded to a surrounding array of molecules or anions. These peripheral species are called ligands or complexing agents. Those atoms within the ligands that are directly bonded to the central metal atom are called

Entombment of Christ painted by Pieter van der Werff in 1709. Prussian blue is used in the sky, and in Mary's veil.

Alfred Werner (1866–1919).

A coordination polymer is an infinite array of coordination complexes in which metal ions are bridged by multidentate ligands. This definition was given by J. C. Bailar in 1964.

donor atoms; typically they are electron rich species like oxygen or nitrogen.

The typical coordination complex is a discrete species. It can be neutral, as for example $[Co(NH_3)_4 Cl_2]$, in which case the charges on the metal ion and the anions are internally balanced. Alternatively, it can be a charged species, like $[Co(H_2O)_6]^{3+}$ in which case electroneutrality is obtained with a compensating ion. However, in both these situations, the coordination complex is a molecular entity. More specifically, its surface is topologically equivalent to a sphere. Alternatively, one can term it as a zero-dimensional object. In a coordination polymer, however, the ligands are *exodentate*. They contain more than one donor atom. Therefore, the ligands are able to connect different metal ions into infinite arrays. In other words, the coordinated units form a repeating structure. Hence these substances are called *polymers*. To summarize, the exodentate ligands, also called spacers or linkers, are coordinated to more than one metal ion to form the polymer. Depending on the number of donor atoms in the ligand and the nature of the metal ion coordination, the coordination polymer is one-dimensional. Alternatively, the metal ions and the ligands form a two-dimensional array; the coordination polymer is two-dimensional. Or else, the metal atoms and the ligands form a three-dimensional array; the coordination polymer is three-dimensional. This last category of coordination polymers are conceptually very similar to extended solids like diamond, ZnS and NaCl.

Let us take a simple example. A metal ion capable of octahedral coordination such as Cu^{2+} or Ni^{2+} will form a coordination complex of the type $[M(py)_2X_4]$ with the two pyridine ligands in *trans* positions. These ligands can be replaced by 4,4'-bipyridine (bpy) to give a linear coordination polymer.

We also need to distinguish between a coordination polymer and solids in which discrete coordination complexes are hydrogen bonded to one another. Lastly, the linker ligands must be organic for the resulting compound to be called a coordination polymer. This excludes from the definition of a coordination polymer, polymeric species formed between metal ions and other inorganic molecules or ions such as halides, oxides, hydroxides, sulfides and so on.

Coordination polymers often precipitate out by mixing the metal ions and organic spacer ligands. They are in general highly crystalline and are insoluble in common solvents. Because of the crystalline nature of coordination polymers and also their relative insolubility, single crystal X-ray crystallography remains the method of choice to determine the structures of these substances. The tremendous growth in the study of coordination polymers is in no large measure due to the ready availability of single crystal X-ray diffractometers during the past decade in many parts of the world. It is interesting to examine the growth of this field. The crystal structures of $Zn(CN)_2$, $Cd(CN)_2$, and the Hofmann clathrate were determined around 70 years ago. However, it was only after Robson's analysis of the $Zn(CN)_2$ structure as an infinite coordination polymer in 1989 that the field really took off. A sustained interest in structural topologies and potential applications led to identification of three-dimensional porous coordination networks as important targets. These compounds are also called metal–organic frameworks (MOFs) and they continue to be studied intensively.

In coordination polymers, the coordination bonds are the strongest interactions used in the assembly of the pre-desired network structure. Other weak intermolecular interactions play a less important role. While defining the dimensionality of a coordination polymer, only the connectivity arising from the coordination bonds should be considered. For example, a one-dimensional coordination polymer may be further hydrogen bonded to give a two-dimensional network. Such a structure should not be classified as a two-dimensional coordination polymer but rather a hydrogen bonded one-dimensional coordination polymer.

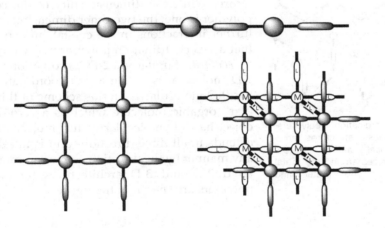

Coordination polymers of different dimensionality.

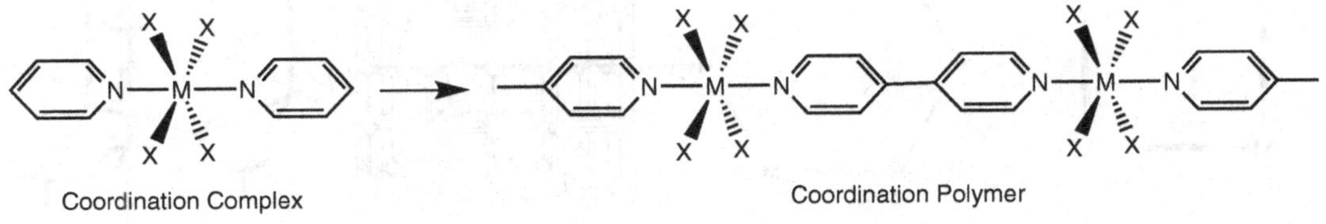

Coordination Complex

Coordination Polymer

From a coordination complex to a coordination polymer. X is a terminal ligand, anion or solvent

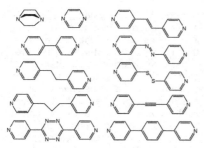

Some commonly used neutral bidentate linkers.

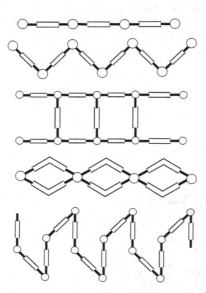

Some   1D   coordination   polymers.
Variations   in   these   structures   arise
from   geometrical   and/or   stereo-
chemical differences around the metal
centers   and   in   the   metal-ligand
arrangements.

Crystal engineering is concerned with the study and understanding of intermolecular interactions in the context of crystal packing, in the design strategies geared towards obtaining particular packing arrangements that are associated with pre-desired properties. Coordination polymers fulfill these aims and characteristics of crystal engineering. The coordination bonds are the interactions of choice in crystal design. Design strategies are generally retrosynthetic in nature and the connectivity and dimensionality of the coordination polymers obtained depend on the metal/ligand ratio, linker functionality and the geometry and coordination number (between 2 and 10) of the metal center. A large number of properties are currently sought after today, most notably gas absorption in open architectures.

7.2   Classification Schemes

It is logical to classify coordination polymers based on their dimensionality. Some one-dimensional (1-D) coordination polymers are shown in the accompanying figure. While the dimensionality of the coordination polymer is one, the two other dimensions are open for further interactions in the crystal, and so we can say that a 1-D coordination polymer has two-dimensional *molecularity*. Similarly, a 2-D coordination polymer has 1-D molecularity while a 3-D coordination polymer has 0-D molecularity. In this scheme of things, a "regular" organic molecule which is a zero-dimensional object has 3-D molecularity. Its "molecular" behavior extends in all three directions and is not delimited in any manner by the direction. Some commonly encountered 2-D and 3-D architectures for coordination polymers are shown in the figure.

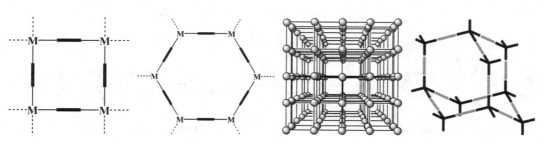

Some 2-D and 3-D architectures.

7.3   Crystal Design Strategies

The M–L coordination bond (40–120 kcal mol^{-1}) is of an energy that is strong enough to confer structural stability and yet weak enough that it can be broken easily under ambient conditions in solution. This latter attribute is of significance in correcting improperly assembled components, or errors, in the assembly of a coordination polymer during crystallization. These compounds illustrate a nice balance between kinetic accessibility and thermodynamic stability. Therefore, a periodic 3-D structure with long range order can be achieved. In contrast, this is not usually possible for covalently bonded organic polymers, in which the C–C bonds (~100 kcal mol^{-1}) once formed cannot be broken and the mistakes in polymerization once made are locked in, resulting in a material with much less periodic ordering, that is crystallinity. It is this long range ordering in coordination polymers enables detailed structural determination through X-ray crystallography. The coordinate bond is also directional, with predictable geometries around the metal center, allowing design to be attempted with some degree of confidence.

The design strategy for the synthesis of coordination polymers is concerned with a particular structural topology, an expected functional property of the resultant bulk material or both. Successful incorporation of optical, chiral, catalytic, electrical and magnetic properties, or porosity for inclusion of solvents and gases within the components of the coordination framework array is an integral part of the design strategies, which also must take into account the metal to ligand ratio, coordination environment, oxidation states and electronic properties of the metal ions, and the shape, geometry and functional groups present in the organic spacer ligands.

7.4   Network Topologies

It is convenient to simplify the description of coordination polymers in terms of networks in well-known inorganic structures. A net consists of nodes and node connectors. For any network depiction of a crystal structure to be of practical utility, only the strongest interactions in the crystal should be chosen as node connectors. In Chapter 3 we saw how the tecton-synthon model is used in analyzing crystal structures that are held together with moderately strong hydrogen bonds.

Metal clusters can be used instead of metal ions for assembling 3-D porous coordination polymeric networks (metal-organic frameworks, MOFs). The pore size can be expanded with a longer spacer ligand such as biphenyl, tetrahydropyrene, pyrene, or terphenyl without changing the framework topology. For example, reaction of $Zn(NO_3)_2 \cdot 4H_2O$ with various linear dicarboxylate spacers gives octahedral metal clusters $Zn_4O(O_2C\text{-})_6$ also called secondary building units or SBU. The SBUs are connected with organic spacers to form 3-D MOFs with primitive cubic topology.

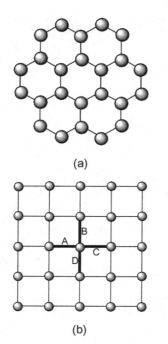

(a)

(b)

(a) Graphite (6,3) net (b) Square (4,4) net showing links.

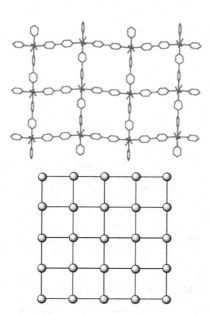

A coordination polymer with square grid structure and its topological representation. [Cd(4,4'-bpy)$_2$(NO$_3$)$_2$].

While the synthon in an organic crystal is a structural unit composed of molecular fragments and the weak interactions that connect them, the synthon in a coordination polymer is nothing but the coordination bond itself. So, the description of connectivity patterns in a coordination polymer is more straightforward than in an organic crystal structure.

Crystal structures of inorganic compounds have traditionally been described in terms of nets in which the atoms are the vertices and bonds are the links (edges) between them. Zeolites, for example, are composed of SiO$_4$, PO$_4$ and AlO$_4$ tetrahedra joined with –O– links. A. F. Wells compiled and classified a large number of inorganic crystal structures in terms of nets. Robson extended this net-based approach to the design of new coordination polymers. Each linker is connected to two nodes and each node may be linked to three or more linkers. Many infinite nets can be related to well-known mineral structures. Let us now familiarize ourselves with the net symbols and their nomenclature.

7.4.1   *Net Symbols and Nomenclature*

The notation (n,p) was used by Wells to represent the topology or connectivity of a given net, where n is the number of nodes in the smallest closed circuit (also called fundamental ring) in the net and p is the number of links from each node to neighboring nodes. Thus, graphite is a (6,3) net. The number 6 in this symbol indicates that the smallest circuits in the net are hexagons, and the number 3 indicates that each node is connected to three other nodes. Similarly diamond is represented as a (6,4) net.

The Wells point symbol is often used to describe more complex nets. Nodes are connected with links. A ring is the shortest circuit that brings one back to a starting node. The point symbol takes the form $A^a.B^b.C^c..$ in which the letters A, B, C... denote the sizes of the rings originating from a node. The superscripts a, b, c, ... are the numbers of circuits at that node. For example, if we consider the (4,4) net, there are four links for a given node namely A, B, C and D. For the four link pairs (AB, BC, CD and AD), the fundamental rings contain four nodes each. On the other hand, the shortest path connecting the AC or BD link pairs has six nodes. Since this is not the shortest ring, it is called a circuit. Hence the point symbol is $4^4.6^2$.

We now analyze the compound [Cd(4,4'-bpy)$_2$ (NO$_3$)$_2$] with respect to connectivity. For each Cd(II) ion, the N atoms from four different bpy ligands occupy the equatorial positions while the axial positions are occupied by the monodentate nitrate ligands. Each 4,4'-bpy ligand bridges two different Cd(II) centers, resulting in a molecular square that is extended in two-dimensions leading to a square grid network. This can be conveniently written as just square grids. It is obvious that Cd(II) can be replaced by other octahedral metal ions having this square planar connectivity or by a metal ion with just square planar geometry. Similarly any of the linear spacers shown in page 3 can be used instead of bpy. The Wells symbol is (4,4).

Geometrical information is not revealed in a topological description. For example, a hexagonal network made up of triangular 3-connecting nodes and a brick-wall type network make up of T-shaped 3-connecting nodes are geometrically different, yet both are topologically identical and are referred to as a (6,3) net. [Cu$_2$(pyz)$_3$] SiF$_6$ has honeycomb structure in which Cu(I) has a 3-connected node with trigonal planar geometry and bridged with pyrazine. Further, the overall connectivity may be highly distorted or the distance between the nodes may vary in the same structure, but they will still have the same topology since it is the connectivity that defines the topology and not the geometry of the nodes.

7.4.2 Topologies of Three-dimensional Structures

Some well-known 3-D nets named after common inorganic structures are now discussed. They are: (i) the diamond, Lonsdaleite, quartz, feldspar and zeolite related nets with 4-connecting, tetrahedral centers; (ii) the NaCl net (with 6-connecting octahedral centers); (iii) the NbO and CdS nets involving square planar 4-connecting centers; (iv) the PtS net (with equal numbers of tetrahedral and square planar centers).

7.4.2.1 Diamond Topology

One of the most frequently encountered three-dimensional net is based on diamond (**dia**). The structure of diamond is made up of tetrahedral sp^3 carbon atoms. The structures of Si, Ge, ZnS, ZnSe, GaAs, CdS, CdSe, CuInS$_2$ and CuInSe$_2$, have the diamond architecture. This topology does not have inversion

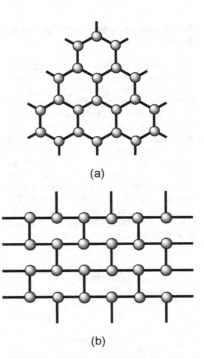

(a)

(b)

Two geometrically different but topologically identical nets (a) Honeycomb (b) Brick wall.

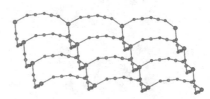

The nodal geometry need not be the same as the geometry of the metal ions. Sheet structure of the square planar (4,4) net formed by tetrahedral Zn(II) and the angular spacer ligand dicyanamide, N(CN)$_2^-$.

This website contains searchable 3D nets: Reticular Chemistry Structure Resource http://rcsr.anu.edu.au./home

Recently, a number of computer programs have been developed to aid in the identification of nets and also to calculate various mathematical descriptors of nets. These include Systre, OLEX and TOPOS.

Metal ions as linkers and ligands as nodes

When metal centers are connected by bidentate spacer ligands in 1:1 ratio, either the metal or the ligand can be termed as a node. Linear, zigzag and helical coordination polymers cannot be assigned a Wells symbol. Similarly if the organic ligand is connected to more than two metal ions, it becomes a node. For example, if all three carboxylate groups in 1,3,5-benzene tricarboxylate (BTC) are connected to metal ions, BTC becomes a node. If it is connected only to two metal ions, then it is a linker. The reader is asked to sketch these possibilities.

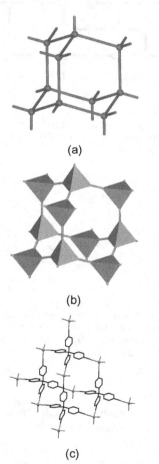

Diamond net (a) $Zn(CN)_2$ structure (b) $Zn(CN)_2$ topology (c) $[Cu\{C(C_6H_4-CN)\}_4]^+ \cdot BF_4 \cdot xC_6H_5NO_2$ without anions or solvent.

centers at the nodes and hence these structures are often found to crystallize in noncentrosymmetric space groups exhibiting nonlinear optical properties. (Section 7.9). We have referred to Robson's description of $Zn(CN)_2$ as the starting point for the field of coordination polymers. In this compound, the cyanide anions bridge the tetrahedral metal centers. In this anion, both the C- and N-atoms can bond coordinatively to Zn(II). In $Zn(CN)_2$, the Zn(II) ion is linked to two C-atoms and to two N-atoms giving a diamond topology. The Zn···Zn distance in this structure is 5.13 Å as compared to the C–C distance of 1.54 Å in diamond.

Similarly one can also construct a diamond network by selecting a tetradentate ligand with tetrahedral geometry such as *tetrakis*(4-cyanophenyl)methane and use it as a linker for tetrahedral metal ions. Indeed $[Cu\{C(C_6H_4\text{-}CN)_4\}] BF_4 \cdot xC_6H_5NO_2$ forms a diamondoid structure. The same network is provided by $[Cu(4,4'\text{-}bpy)_2]PF_6$ and by $[Cu(2,5\text{-dimethylpyrazine})_2]PF_6$.

Interestingly, the quartz topology (**qtz,**) has been observed for $[ZnAu_2(CN)_4]$. Here the $[Au(CN)_2]^-$ anions act as rod linkers for the tetrahedral Zn(II) nodes. Lonsdaleite or hexagonal diamond is another diamondoid net and a form of SiO_2 called cristobalite has this topology. Let us consider $Cd(CN)_2$. The unsolvated structure has two interpenetrating diamond nets (Section 7.5), while the solvated $Cd(CN)_2 \cdot CCl_4$ has only one diamond net. In contrast, $Cd(CN)_2 \cdot 0.5(n\text{-}Bu_2O \cdot H_2O)$ has the Lonsdaelite topology. Solvents used for crystallization can have a great influence on the overall topology. Still, the diamond net is almost the default topology for tetrahedral nodes.

7.4.2.2 NaCl Topology

The simplest six-connected net with octahedral nodes has the primitive cubic (**pcu**) structure. This is illustrated by the α-Po crystal structure. The same octahedral topology is also seen in NaCl and ReO_3. In NaCl, each Na^+ is linked octahedrally to six Cl^- anions, and in ReO_3 each Re^{6+} is linked octahedrally to six O^{2-} anions. All these structures are said to have the NaCl topology. The structure of Prussian Blue, $Fe^{III}_4[Fe^{II}(CN)_6]_3 \cdot xH_2O$ belongs to this net. Here each Fe is an octahedral node bridged by six CN^- linkers. $[Ag(pyz)_3]SbF_6$ is composed of Ag(I) metal ions unusually bonded octahedrally to six nitrogen atoms from six different pyrazine spacers. Six-connected octahedral nodes can also be constructed with [Zn$_4$O(O$_2$CR)$_3$] and [Zn$_3$(O$_2$CR)$_6$] clusters. The structures of [Zn$_4$O(1,4-bdc)$_3$]·8DMF·C$_6$H$_5$Cl and [Zn$_3$(1,4-bdc)$_6$]·6MeOH illustrate more complex variations of the **pcu** topology.

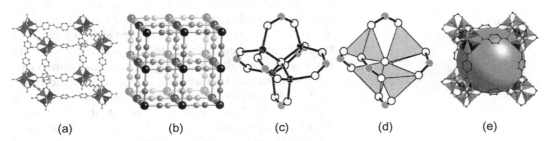

(a) A **pcu** net (b) Prussian Blue (c)-(e) [Zn$_4$O(1,4-bdc)$_3$] showing the building blocks.

7.4.2.3   *NbO and CdSO$_4$ Topologies*

A variety of 4-connected nets can be constructed using square planar nodes. Of these the most common are the **nbo** and **cds** nets with NbO and CdSO$_4$ structures. The **nbo** net is the second most frequently found net after diamond in coordination polymers. This net is built from square planar nodes that are connected to each other with a 90° twist. [Cu$_2$(2-Br-BDC)$_2$(H$_2$O)$_2$].8DMF.2H$_2$O, furnishes a good example of this net; here the Cu$_2$(O$_2$C-)$_2$ paddle-wheel units act as square nodes. They are joined by the aromatic linkers.

Both **nbo** and quartz nets have point symbol 6^4.8^2 and hence they are designated specifically as 6^4.8^2-*a* and 6^4.8^2-*b*. In the **cds** topology only half of the adjacent nodes are perpendicular and the rest are coplanar. This **cds** net is found in SmL$_2$(NO$_3$)$_3$·0.5H$_2$O, where L = 4,4'-bipyridine-*N,N'*-dioxide.

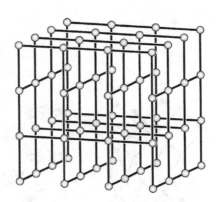

Topology of the **cds** net.

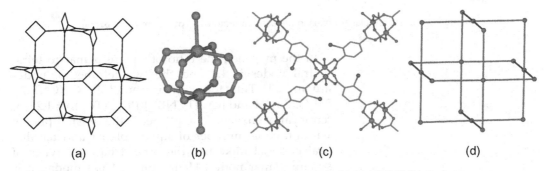

(a) NbO topology (b), (c) and (d) show the [Cu$_2$(2-Br-BDC)$_2$(H$_2$O)$_2$].8DMF.2H$_2$O structure.

7.4.2.4   *PtS and Related Topologies*

So far, the nets we have studied contain only a single type of node, in other words they are *uninodal*. If the same structure contains more than one type of node, interesting possibilities arise. When square planar and tetrahedral nodes are linked, three *binodal* structures are possible. They are **pts** (platinum sulfide), **ptt** (PtS twisted) and **mog** (moganite). The structure of **pts** (cooperite mineral) is made up of an equal number of square planar and tetrahedral nodes. The connectivity in the following coordination polymers adopts the **pts** topology: $(Me_4N)[Cu^I\{Pt(CN)_4\}]$, $[Cu^I(tcnb)]PF_6$ (tcnb = 1,2,4,5-tetracyanobenzene), $[Cu^I\{Cu^{II}(tpyp)\}]BF_4$ (tpyp = 5,10,15,20-tetra(4-pyridyl)porphyrin), and $[Cu_2(atc)_2]$ (atc = adamantane tetracarboxylate). In the first three compounds Cu(I) readily forms tetrahedral nodes while in the last case, the square planar nodes formed by "$Cu_2(O_2C-)_4$ paddle-wheel" are connected to tetrahedral linkers or struts. In $[Co(rtct\text{-}tpcb)(F)_2]\cdot5H_2O$ (*rtct*-tpcb = *regio trans, cis, trans*-tetrakis(4-pyridyl)cyclobutane), the square planar Co(II) nodes are attached to tetrahedral *rtct*-tpcb struts.

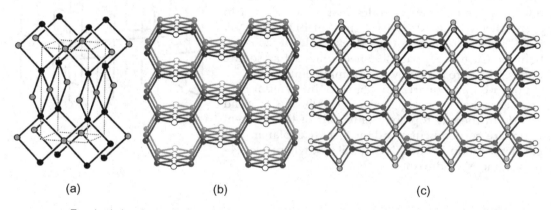

(a)   (b)   (c)

Tetrahedral and square planar nodes in the same structure. (a) **pts** (b) **mog** (c) **ptt** nets.

In the **mog** net, the ratio of square planar to tetrahedral nodes is 1:2 and the two tetrahedral nodes are linked. This topology is found in $Cd(CN)_2\cdot2/3H_2O\cdot^tBuOH$ and $[Cu_2(TCNB)_3](PF_6)_2$ (TCNB = 1,2,4,5-tetracyanobenzene). The **ptt** net is the twisted **pts** net with an equal number of square planar and tetrahedral nodes. Unlike **pts**, the **ptt** net has two types of square planar nodes. Hence this net is trinodal; it is seen in $[Zn(rtct\text{-}tpcb)(H_2O)_2](ClO_4)_2\cdot6.5H_2O$ wherein the tetrahedral *rtct*-tpcb struts are connected to square planar Zn(II) nodes.

The retrosynthetic approach is very useful in the design of simple topologies. The diamond topology is an easy synthetic target for 3-D coordination polymers. One can extend this idea to design other 3-D topologies by mixing appropriate nodes. A few designed uninodal and multinodal targets are summarized in the Table.

7.5 **Supramolecular Isomerism**

Supramolecular isomerism in coordination polymers can be understood with relation to two well known concepts: isomerism in coordination complexes and polymorphism in organic crystals (Chapter 5). A coordination complex like $[Co(NH_3)_4Cl_2]$ can have different structures depending on whether the Cl-ligands are *cis* or *trans* in the octahedral arrangement. These structures are isomers. Isomerism in coordination complexes arises because of differences at the molecular level. These include differences in the geometry of the metal ions (structural and geometrical isomerism) or from differences in the ligands (linkage and conformational isomerism) or other factors (ionization, solvation, coordination, optical). Isomerism in coordination complexes is closely related to isomerism in organic molecules (structural and stereo). In all these cases, there are differences in the molecular structures of the isomers.

Polymorphism is the phenomenon wherein the same molecule has different crystal structures. In coordination polymers, repeating building blocks are connected with coordination bonds to give infinite arrays. In many cases, the same building blocks can be connected in different ways and therefore different structures are formed. For example, let us consider the formation of $[Co(dpee)_{1.5}(NO_3)_2]$ (dpee = 1,2-bis (4-pyridyl)ethane) under different crystallization conditions. In a MeOH/MeCN solvent mixture $Co(NO_3)_2$ and dpee in a 1:1.5 ratio give a linear chain made up of rings. When a MeOH/CHCl$_3$ solvent mixture is used, a ladder-like structure is obtained. The same solvent, when contaminated with ferrocene, yields a bilayer structure. In all the three polymeric structures, the same metal to ligand stoichiometry is maintained. The metal center, in all cases, is pentagonal bipyramidal with a T-shaped arrangement of pyridyl groups and two chelating nitrate ligands.

Some common nets

Node Geometry	Node Geometry	Topology
Triangle	Square	Pt_3O_4
Triangle	Tetrahedron	Boracite
Triangle	Octahedron	Rutile
Square	Square	NbO
Tetrahedron	Tetrahedron	Diamond
Square	Tetrahedron	PtS
Tetrahedron	Octahedron	Corundum
Tetrahedron	Cube	CaF_2
Octahedron	Octahedron	NaCl

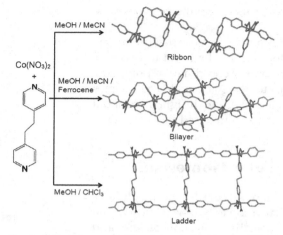

Connectivity between Co(II) and dpee in [Co(dpee)$_{1.5}$(NO$_3$)$_2$].

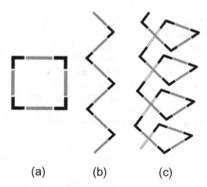

Three supramolecular isomers observed for ML compounds (a) Ring (b) Zigzag (c) Helical structures.

How does one describe these structural variations? One could call them polymorphs but how does one define the "molecule"? Remember that in a pair of polymorphs, the "same" molecule takes "different" crystal structures. Are the "molecules" the "same" in the [Co(dpee)$_{1.5}$(NO$_3$)$_2$] structures? These difficulties arise from the fact that the concept of a "molecule" is very precisely defined in organic chemistry but this is not so in the field of coordination polymers, where the concept of dimensionality of molecularity also needs to be considered.

In these circumstances, Zaworotko suggested that the well-known concept of molecular isomerism in coordination complexes can be extended to coordination polymers. He suggested the term *supramolecular isomer* to describe the kinds of variation seen in the [Co(dpee)$_{1.5}$(NO$_3$)$_2$] structures. This is a useful concept. Supramolecular isomerism is the existence of more than one type of network superstructure for the same molecular building blocks.

A few examples for supramolecular isomerism are now described for different coordination polymers. These are classified based on the metal to ligand ratio.

1. ML compounds. When the metal has a *cis* geometry and the spacer is linear, three possibilities arise for supramolecular isomerism. These are the ring, the zigzag chain and the helix. The *trans* geometry will produce only the linear chains. Occurrence of supramolecular isomers for coordination polymers with a 1:1 M:L ratio is quite common and is obtained when the experimental conditions are varied slightly.

2. $ML_{1.5}$ compounds. The possible structures with linear spacers include the ladder, the parquet floor (also called herringbone), the brick wall, the honeycomb (also called hexagonal sheet) and the bilayer. The geometry of the nodes can be either T-shaped or trigonal planar for linear spacer ligands. Some of these possibilities have already been described for $[Co(dpee)_{1.5}(NO_3)_2]$.

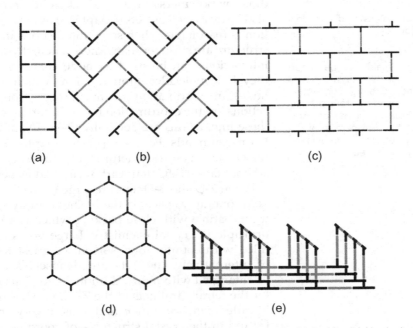

Five supramolecular isomers observed for $ML_{1.5}$ compounds. (a) Ladder (b) Herringbone (c) Brick wall (d) Honeycomb (e) Bilayer.

3. ML_2 compounds. The possibilities with linear and angular bidentate ligands are the ribbon and necklace polymers (1-D), the square grid (2-D) and the diamond topology (3-D). The square grid is easily

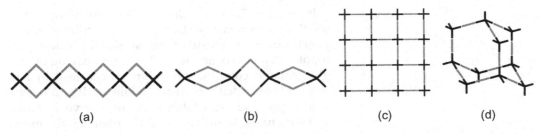

Schematic representation of supramolecular isomers formed by ML_2 coordination polymer. (a) Ribbon (b) Necklace (c) Square grid and (d) Diamondoid structures.

Interpenetration of two diamondoid networks in Cd(CN)$_2$.

A network is said to be inter-penetrating, when two or more nets which are not directly connected, cannot be separated topologically without the breaking of bonds. This property holds true irrespective of the dimensionality of the coordination polymer being considered.

predicted and an early example is [Co(pyz)$_2$Cl$_2$] in which the pyz and the Cl$^-$ bridge the square planar Co(II) ions.

7.6   Interpenetration

It is very convenient to describe the structures of coordination polymers as networks. Organic structures can also be described as nets (Chapter 3). A network depiction is reasonable when some interactions in a crystal structure are stronger than others. Then, these stronger interactions can be used as node connections when they link molecules, atoms or ions which act as the nodes. In coordination polymers, these stronger interactions are the coordination bonds. In organic crystals, these interactions are generally O–H···O and N–H···O hydrogen bonds. For example, in Chapter 3, we analyzed the crystal structures of trimesic acid and adamantane-1,3,5,7-tetracarboxylic acid as nets.

When 2-D and 3-D nets are large in size, they create voids or empty spaces in them. These empty spaces are incompatible with close packing which is a universal principle in crystal chemistry. Large voids are either filled with counter ions, solvent or guest molecules, or alternatively, the nets are themselves entangled or entangled with more additional nets. This is shown for the tetrahedral nets of the Zn(CN)$_2$ structure. This is called *interpenetration* and it is a very important feature in the crystal structures of coordination polymers. When only two networks are interpenetrated, as in Zn(CN)$_2$, the network is described as being twofold interpenetrated. As the M–L–M distance becomes larger, the degree of interpenetration increases. More generally, networks can be *n*-fold interpenetrated with values of *n* up to 12 being known. The degree of interpenetration in adamantane-1,3,5,7-tetracarboxylic acid is 5.

In Section 7.4.1 we showed how to identify the topology of a network structure. To describe an interpenetrated network structure, we should include the topology of interpenetration. The interweaving of single stranded coordination polymers can lead to 1-D and 2-D entangled structures, while ladder and step-ladder polymers can interpenetrate to give 2-D and 3-D structures. In the case of 2-D polymers, the mean planes of the nets can interpenetrate either in a parallel or an inclined fashion.

Four distinct types of 2-D entanglement are possible. Parallel interpenetration can occur only if the 2-D sheets are highly corrugated, as in [Ag(dps)$_2$]PF$_6$ (dps = 4,4'-dipyridylsulfide. Inclined interpenetration can lead only to 3-D entanglements. This can be divided into parallel/parallel, diagonal/diagonal and parallel/diagonal inclined interpenetrations. [Co(4,4'-bpy)(dca)]·½H$_2$O·½MeOH) (dca = dicyanamide) and [M(4,4'-bpy)$_2$(H$_2$O)$_2$]SiF$_6$ (M = Zn, Cd, Cu) are examples of parallel/parallel and diagonal/diagonal inclined interpenetrated structures.

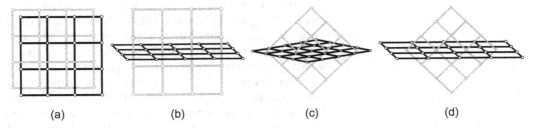

(a)   (b)   (c)   (d)

Interpenetration topologies of some 2-D (4,4) nets: (a) Parallel interpenetration (b), (c), (d) Inclined interpenetration: (b) Parallel/parallel (c) Diagonal/diagonal (d) Parallel/diagonal.

Among 3-D structures, interpenetration in **dia** and **pcu** nets is noteworthy. We have already referred to the double interpenetration in Zn(CN)$_2$. 12-fold interpenetration is seen in [Cu(SO$_4$)(L)(H$_2$O)$_2$] (L = *N,N'*-di (4-pyridyl)adipoamide). Further, the compounds [Zn$_4$O{(O$_2$C)$_2$R)$_3$}] (where (O$_2$C)$_2$R = various linear dicarboxylates) all have the **pcu** topology and almost all are doubly interpenetrated.

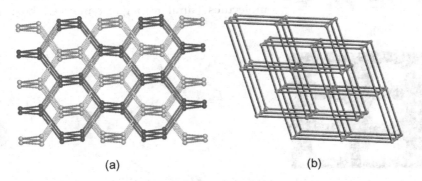

(a)   (b)

Double interpenetration in the (a) **dia** and (b) **pcu** nets.

The lengths of the spacers can be varied to adjust the extent of interpenetration in these structures. For example, the degree of interpenetration increases with

Interpenetration in Ag(NC-(CH$_2$)$_n$-CN)BF$_4$

n	interpenetration
1	2
2	5
4	6
8	8

increasing the lengths of the alkyl chains in the diamondoid framework structure of Ag(NC-(CH$_2$)$_n$-CN)$_2$(BF$_4$). However, the bulkiness and conformation of the ligands, the counter ions, solvents and experimental conditions (such as dilutions) play important roles and have a pronounced influence upon the degree and nature of interpenetration.

7.7   Porous Coordination Polymers

An important property of coordination polymers is porosity. Porous crystalline solids are shown to have potential applications such as gas and liquid storage, separation, ion exchange, heterogeneous catalysis, sensing, nanoreactors and drug delivery. The crystal packing features of coordination polymer networks will determine whether voids, cavities and channels are generated. Interpenetration will reduce the pore sizes and efficiency of gas absorption. A group of porous 3-D coordination polymers that are constituted with metal ions such as Fe(II), Co(II), Cu(II) and Zn(II) and use dicarboxylates as linkers have been classified as metal-organic framework compounds, or MOFs, by Yaghi who has done extensive work in the area of porous solids.

Kitagawa classified porous coordination polymers into three categories. In the Type I PCPs (first generation), the porosity cannot be maintained without the guest molecules and indeed, the framework structure collapses upon the removal of the guest molecules (similar to guest induced host assembly,

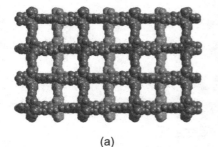

(a)

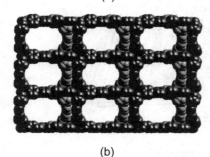

(b)

Packing of 2-D layers of [NiL$_2$(NO$_3$)$_2$] with *o*-xylene and mesitylene. A third generation porous coordination polymer.

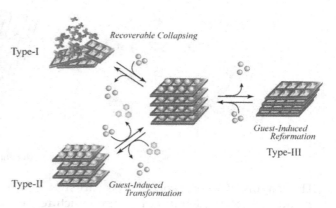

Classification of porous coordination polymers.

Section 6.3). In the Type II (second generation) PCPs, the structures are robust and preserve the porosity even after guest removal. Guest uptake and removal occurs reversibly without the destruction of the framework structure and porosity. The Type III (third generation) framework structures are flexible and dynamic, and respond to external stimuli, such as light, electric field, guest molecules. Further they change their channels and pores reversibly.

First and second generation PCPs are fairly common. An interesting example of a third generation framework compound is illustrated. In the square grid 2-D layer structure [NiL$_2$(NO$_3$)$_2$]·4(*o*-xylene) (L = 4,4'-bis(4-pyridyl)biphenyl), the *o*-xylene guest molecules can be completely exchanged by the bigger molecule mesitylene. This is possible because of a sliding of the adjacent square grid layers leading to bigger channels as shown.

7.7.1 *Pore Size*

One of the major goals in the design and synthesis of coordination polymers is to control the size and shape of the pores. If the framework is cationic, then non-coordinating anions will occupy the pores and this unnecessarily reduces the pore size. Hence it is wise to synthesize neutral frameworks to generate PCPs with increased pore volume. It may look simple and easy to increase the pore size by increasing the length of the spacer ligands, and this strategy indeed works in some cases, as for example the Cu(II) based MOFs illustrated here. However, if the length of the linker is increased beyond a certain limit, interpenetration occurs. If the structure is already interpenetrated, the degree of interpenetration usually increases.

Interpenetration of networks remains a challenge for crystal engineers who seek to make porous coordination polymers. The use of metal clusters as secondary building units (SBU) as nodes instead of single metal ions has been found to reduce this problem. A representative set of compounds is shown here.

7.7.2 *Gas Sorption and Storage*

The most interesting characteristic of porous MOFs is their exceptional specific surface area of 1000–6000 m^2 g^{-1} as compared to conventional zeolites (~500 m^2 g^{-1}). This allows them to trap various gases efficiently.

Properties of porous materials are determined by adsorption of guest molecules on the solid surface. This in turn depends on the nature of the interactions between guest molecules and the surface, pore size and shape. Pores are classified based on their size as given in the table below.

Classification of pores

Pore Type	Pore size (Å)
Ultramicropore	<5
Micropore	5–20
Mesopore	20–500
Macropore	>500

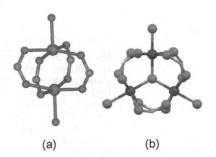

(a)   (b)

(a) 4-connecting SBU with a square planar node. M$_2$(solvent)$_2$(O$_2$CR-)$_4$; (b) 6-connecting SBU with an octahedral node. [Ni$_3$O(O$_2$CR)$_6$(H$_2$O)$_3$]$^{2-}$ (R = 9, 10-anthracenyl group).

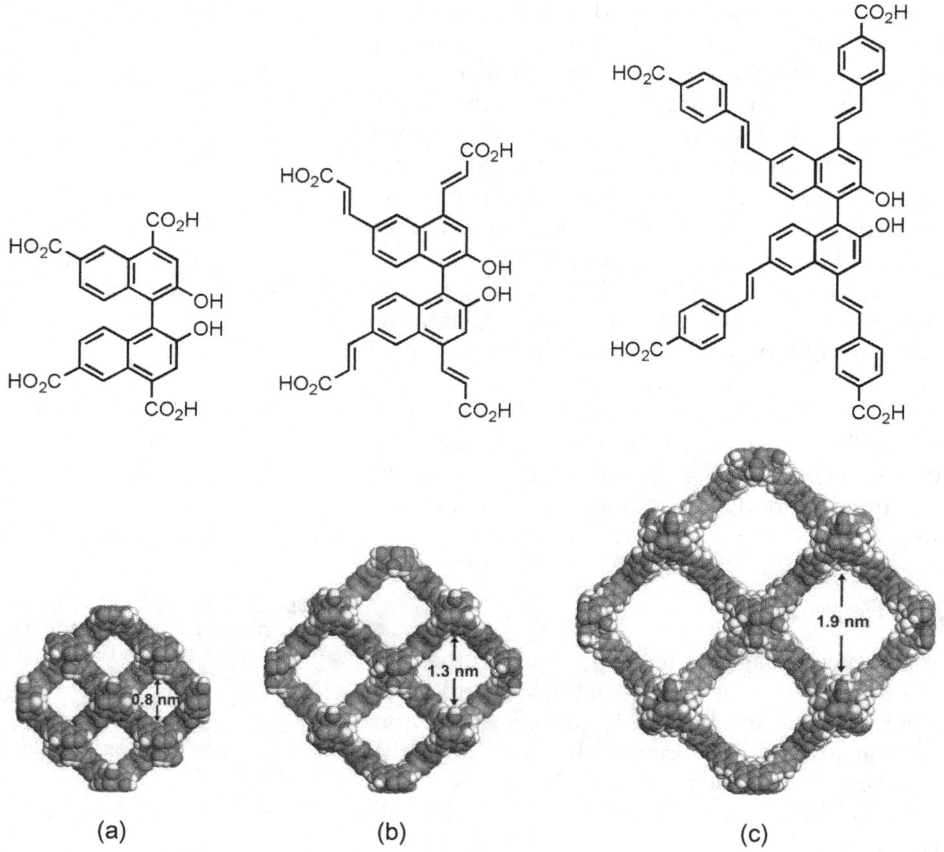

(a)   (b)   (c)

Effect of ligand size on pore size in a series of MOFs based on Cu(II).

**US Department of Energy targets for on-board hydrogen
storage systems**

Storage Parameters	Units	2010	2015	Ultimate
System Gravimetric Capacity (net useful energy/max. system mass)	kg H_2/kg system	0.045	0.055	0.075
System volumetric Capacity (net useful energy/max. system volume)	kg H_2/L system	0.028	0.040	0.070
Min. /max. delivery temperature	°C	−40/85	−40/85	−40/85
Cycle life (1/4 tank to fill)	cycles	1000	1500	1500
Max. delivery pressure from storage system	atm (abs)	100	100	100
System fill time (for 5 kg H_2)	min	4.2	3.3	2.5

Among these, H_2, and CH_4 are significant in the context of energy storage requirements. MOFs may also help to capture CO_2 emissions which are a major cause of global warming and other environmental hazards. The ultimate goal for practical applications of MOFs is to store significant amounts of H_2 at ambient temperatures and closer to atmospheric pressure. The details of the goals set by the US Department of Energy are shown in the Table.

Weak dispersive interactions (physisorption) and also stronger chemical associations (chemisorption) are responsible for H_2 binding to surfaces. A larger surface area rather than a larger pore volume favors maximum H_2 uptake by physisorption. This is because H_2 molecules in the middle of the larger pore will have no attraction from the potential surface of the pore walls. Due to this, low density MOFs will have low volumetric H_2 uptake capacity. This can be in principle improved with interpenetration. For example, the doubly interpenetrated $Cu_3(tatb)_2$ ($tatb^{3-}$ = 4,4',4"-*sym*-triazine-2,4,6-triyltribenzoate) was found to adsorb 1.9 wt% of H_2 at 1 bar and 77 K which shows ~0.6 wt% improvement over the non-interpenetrated compound. Further, the absorption CO_2 capacity of 33.5 mmol g^{-1} for $[Zn_4O(BTB)_2]$ (BTB = benzene-1,3,5-tribenzoate) which has a high Langmuir surface area of 5640 $m^2\,g^{-1}$ exceeded the capacities of zeolites and activated carbons. MOFs assembled from imidazolate ligands and their derivatives, called zeolitic imidazolate frameworks (ZIFs) have high thermal and chemical stabilities and show exceptional selective uptake of CO_2 in the presence of CH_4, CO and N_2.

Many porous frameworks are stable after guest removal and the space inside the pore can be used as a reaction vessel. Since the ligand spacers inside the surface of MOFs are also exposed and accessible, it may be possible to derivatize the preinstalled functional groups attached to these spacers without affecting the overall framework integrity of the structures. Such postsynthetic modification of the MOFs has the advantage of introducing diverse range of new functional groups in topologically identical MOFs. Using this method, it is possible to fine-tune and optimize the size and shape of the pore surface of MOF. But the MOF should be sufficiently porous to allow the access of the reagents and stable to the reaction conditions and byproducts formed.

Changes in geometry and local electronic environments due to reversible guest exchange can be used to influence SCO properties drastically without disturbing the single crystal nature of coordination polymers. One such example is illustrated below.

[Fe(azpy)$_2$(NCS)$_2$] · (solvent) (azpy = 4,4'-azopyridine) consists of doubly interpenetrated two-dimensional (4,4) grids, where the interpenetration generates one-dimensional channels that run parallel to the crystallographic *c*-axis. This coordination polymer displays reversible uptake and release of guest. Removal of ethanol from the crystal causes major structural changes such as a rotation of azpy and thiocyanate ligands, a translation of the interpenetrated layers, and the loss of hydrogen bonding interactions. This is accompanied by a change in the space group from monoclinic *C2/c* to orthorhombic *Ibam*. In the monoclinic ethanolate there are two crystallographically distinct Fe(II) centers, of which one participates in a host–guest hydrogen bond. Interestingly, SCO is observed for this Fe(II) between 300 and 150 K due to all Fe(II) sites being in the high-spin state with S = 2 to S = 0. For the desolvated orthorhombic crystal all the iron(II) centers are equivalent and the SCO functionality is lost.

7.8   Properties and Applications

Coordination polymers have a large and varied set of uses, other than their gas sorption applications. These applications exploit the physical and chemical properties of these compounds. The ultimate aim and driving force for crystal engineering is to target useful properties that can be achieved by a certain packing of molecules. This leads to the question about the range of properties that can be targeted and what are the advantages of using coordination polymers for this purpose. In this section some of the properties of the coordination polymers are briefly discussed.

7.8.1   *Magnetism, Magnetic Ordering and Spin Crossover*

Magnetism is a bulk property dependent on strong interactions in all three directions. Strong coordination bonds through bridging atoms between metal ions are much more efficient than interactions through space for communicating magnetic information effectively in coordination complexes. Hence there is great interest in coordination polymers containing metal ions with unpaired *d* and *f* electrons. The magnetic properties of coordination polymers are divided into those display long-range ordering and those that display a spin crossover (SCO). These phenomena involve the ordering of electrons through an external stimulus and are very important in data storage and electronics.

The communication between the spin centers is more effective if a smaller number of atoms bridge the metal ions. Hence it is not surprising that the critical ordering temperatures for the magnetic metals of the first row transition metals is very high and range between 560 and 1380 K. While for oxides (one-atom bridge) of the same metals it does not exceed 900 K and for compounds with two-atom bridges (e.g., cyanide) it is not more than ~350 K. Increasing the number of atoms bridging the metal ions to three (dicyanamide, azide and formate) decreases the transition temperatures to <50 K. One way to increase the T$_c$ is by increasing the dimensionality of the coordination polymers.

The Prussian Blue family of complexes with two atom cyanide bridging has been explored extensively and found to show magnetization at T$_c$ ~375 K. The

Magnetism and Molecular Crystals

Different types of magnetism may be distinguished: diamagnetism, paramagnetism, antiferromagnetism, ferrimagnetism and ferromagnetism. These varieties arise from the diverse ways in which magnetic moments of electrons in molecular and supramolecular arrays can be coupled. Magnetic moments align in different ways and this affects the susceptibility. The moments of the individual centers may be randomly aligned, giving rise to *paramagnetism*. Paramagnetic substances have a low susceptibility and are attracted to external magnetic fields weakly. *Ferromagnetism* is achieved when all the moments are aligned. The susceptibility is very high. *Antiferromagnetism* is obtained when the individual magnetic moments are nearly all aligned antiparallel, and thus nearly cancel each other. This leads to very low magnetic susceptibility. When the substance has no unpaired electrons it is said to be *diamagnetic*. Since there is no contribution to magnetism provided by electron spin ($S = 0$), the magnetic moment is zero. Diamagnetic susceptibility is too small to be measured. Among these varieties, only ferromagnets are important for applications such as quantum computing and molecule-based data storage. The magnetic properties of a molecular material are strongly dependent on the way the molecular building blocks are interconnected and interact. This section highlights how noncovalent interactions can be tuned to control the dimensionality of a magnetic lattice.

Dante Gatteschi pioneered the use of the metal-radical approach to assemble extended lattices. Stable nitronyl nitroxide-based radicals with N-O groups with one unpaired electron in a π^* orbital are used as ligands. Such molecules can be engineered to pack in different ways to control the magnetic interactions.

The mechanism of spin-coupling in molecular magnets differs from that in metals like iron or cobalt. Two distinct mechanisms can be recognized: direct exchange, and indirect or superexchange. Ligand-mediated superexchange interactions in bimetallic complexes can be exploited to engineer magnetic materials. In dimeric complexes, the spin gets paired even if we form a polymeric chain. However, if we could engineer a solid wherein alternate spin centers are occupied by different metal ions with unpaired electrons, one obtains a ferromagnetic crystal. For example, in [MnCu(pba)(H$_2$O)$_3$]·2H$_2$O alternate chains are formed by Mn and Cu ions. It was, however, realized that these materials are antiferromagnetic due to inter-chain interactions. To overcome this problem, the bridging ligand was replaced by another one which prohibits such an interaction resulting in ferromagnetic behavior.

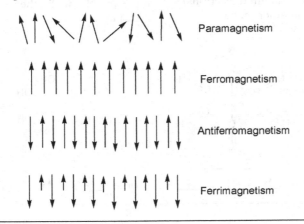

Paramagnetism

Ferromagnetism

Antiferromagnetism

Ferrimagnetism

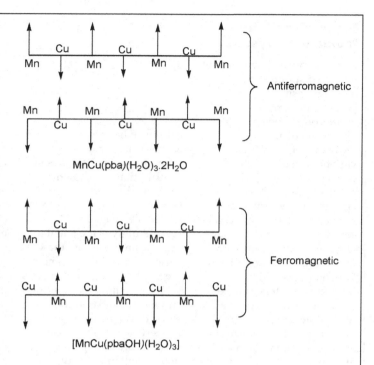

MnCu(pba)(H$_2$O)$_3$.2H$_2$O

Antiferromagnetic

[MnCu(pbaOH)(H$_2$O)$_3$]

Ferromagnetic

Single Molecule Magnet

A *single molecule magnet* (SMM) is a molecule that can be readily magnetized in a magnetic field. It will remain magnetized even after switching off the magnetic field. Magnetism is an intrinsic property of the molecule. No interaction between molecules is necessary for this phenomenon to occur. This makes single molecule magnets fundamentally different from the traditional bulk magnets discussed above. It will show the property even if it is dissolved in a solvent or put in another matrix, like a polymer. A significant feature is that unlike conventional bulk magnets, SMM do not require long-range ordering of spins.

Engineering Molecules with High Spin

The Prussian blue salt, Fe^{III}$_4$[Fe^{II}(CN)$_6$]$_3$·15H$_2$O is ferromagnetic but at very low temperature (T$_c$ is

~5–6 K); this is mainly because the spin carriers Fe^{III} are far away from each other. However, Prussian blue analogs that show varying magnetic behavior can be engineered by exploiting acid-base reactions between say [Cr(CN)$_6$]$^{3-}$ with S = 3/2 (three unpaired electrons) and another metal

complex, say NiL^{2+} (two unpaired electrons and hence S = 1). Crystallization leads to a cubic complex, CrNi$_6$. Since Cr and Ni orbitals are orthogonal, the total spin adds to 15/2. Similar solids can be obtained with different magnetic behavior by replacing Ni^{II} with other metal ions.

Cr^{III} d^3

Ni^{II} d^8

2-D and 3-D networks of oxalate dianion and the azide anion also display ferro- and ferri- long-range ordering. The limitation due to the length of the bridging atoms can be circumvented by using radical organic ligands in coordination polymers. This introduces the possibility of using longer ligands in the framework structure for magnetic communication.

Change in the coordination geometry at the metal centers can cause not only solvatochromism but also solvatomagnetism due to adsorption and desorption of water molecules bonded to the metal ions. For example, the pink $[Co_3\{Cr(CN)_6\}_2]\cdot xH_2O$ shows ferromagnetic coupling with $T_c = 28$ K when the humidity level is 80%. This can be transferred into a blue antiferromagnetically coupled magnet with a $T_c = 22$ K when the humidity level is decreased to 3%. Removal of the aqaua ligands causes a change in geometry from pseudo-octahedral to pseudo-tetrahedral and this in turn leads to a change in the magnetic property.

Cooperativity in the spin crossover (SCO) phenomenon between metal–ligand centers is controlled by the intermolecular lattice effects. If SCO centers are incorporated into the coordination polymers such cooperativity is expected to proliferate through the framework more efficiently than in discrete complexes. The guest molecule interactions with the framework may perturb SCO; thus these PCPs can be exploited for molecular sensing device applications.

7.8.2 *Luminescence and Sensing*

Coordination polymers including MOFs exhibit a wide range of emissive phenomena. They include: (i). Metal-based luminescence; (ii) Ligand-based emission, including ligand-localized emission as well as ligand-to-metal charge transfer (LMCT) and metal-to-ligand charge transfer (MLCT); (iii) Excimer and exciplex emission; (iv) Antennae effects; (v) Adsorbate-based emission and sensitization; (vi) Scintillation (the emission of light in response to ionizing radiation); (vii) Surface functionalization. Some of these unique properties are not observed in conventional coordination complexes; they may arise in MOFs because of the porosity of these compounds. The pores are usually filled with guest molecules in the MOFs and as a result they are in closer proximity with the luminescent centers. Hence the properties of the emission are expected to be influenced by these guests, leading to wavelength

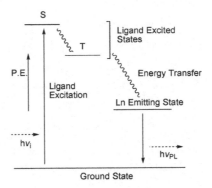

Schematic diagram for the antenna effect.

shifts, intensity changes, or even new emission as a result of excimer or exciplex formation. Increased fluorescence lifetimes and quantum efficiencies may be observed due to the rigidity imposed by the lattice in MOFs that constrains the ligands in ways that are not typically observed for a free complex in solution.

The spacer ligand *trans*-4,4'-stilbene dicarboxylate (sdc) has been used to synthesize $[Zn_4O(sdc)_3]$ and $[Zn_3(sdc)_3(DMF)_2]$ having 3-D and 2-D structures respectively. These two crystals emit purple-blue and deep blue colors respectively. In the 3-D structure the sdc spacers are well separated and have minimal short range cofacial distances resulting in minimal interlumophore interactions. On the contrary shorter $\pi\cdots\pi$ interactions between the sdc spacers yielded a red-shifted emission. Both crystals exhibit remarkable stability towards radiation, making them useful scintillation materials. The photoluminescence (PL) of the dried 3-D structure is red-shifted due to the closer proximity of the sdc ligands upon solvent removal.

There are many advantages in using lanthanide ion MOFs as highly luminescent sensors. The electronic transitions, which give rise to photon emission, are shielded by the *5d* shell and there is little broadening due to solvation. The lanthanide ions most commonly used in sensing applications are Eu(III) and Tb(III) due to their strong visible luminescence in the red and green regions, respectively. A direct energy transfer from the more readily accessed spacer excited state to the appropriate metal energy level is achievable in the presence of strong vibronic coupling between the spacer and metal ions. This coupling leads to a large increase in luminescence and is called the "antenna" effect.

7.8.3 *Nonlinear Optical Properties*

Noncentrosymmetric assembly of molecular building blocks is an essential requirement for a bulk material to exhibit second order nonlinear optical (NLO) properties. Noncentrosymmetry is generally not as prevalent in coordination polymers as it is in organic solids. However, by introducing appropriate unsymmetrical molecular subunits into the backbone of the bridging ligands, it may be possible to assemble noncentrosymmetric solids.

Due to lack of inversion centers, the tetrahedrally connected diamondoid networks are suitable for this purpose. Indeed KD_2PO_4 (the deuterated form of KDP) which has diamondoid topology is noted for its second harmonic generation. But these structures have a tendency to interpenetrate. Regardless of the inherent noncentrosymmetric nature of individual nets, the formation of an even number of interpenetrated nets may lead to inversion. Diamondoid networks with an odd number of interpenetrations constructed using unsymmetrical bridging ligands containing pyridyl and carboxylate groups have been shown to be noncentrosymmetric and exhibit excellent SHG properties. Further, the use of transition metals reduces the optical loss from *d-d* transitions in the visible region.

7.8.4 *Proton Conductivity*

Proton conducting solids find applications in fuel cell technology. The presence of proton conducting carriers in the porous channels of coordination polymers can facilitate the mobility of the proton carriers. For example, in $(NH_4)_2[Zn_2(ox)_3]\cdot adp\cdot 3H_2O$ (ox = oxalate, adp = adipic acid), the anion has the two-dimensional honeycomb sheet structure with NH_4^+ cations and neutral adipic acid molecules in the layers. This coordination polymer exhibits a superprotonic conductivity of 10^{-2} S cm^{-1} at ambient temperature. However, the one-dimensional channels in $[Al(\mu_2-OH)L]$ (where L = 1,4-benzenedicarboxylate, 1,4-naphthalenedicarboxylate) facilitate proton conduction in an anhydrous atmosphere with imidazole being the proton carrier.

3-D coordination polymers (MOFs) possess properties to be excellent electrode materials for lithium ion batteries. The channels in MOFs can be used as hosts for intercalation/de-intercalation of Li-ion so that the parent structure can be retained on cycling. They usually have high thermal stability and in general have insolubility in the electrolytes. However, high electronic and ionic conductivity of electrode materials are necessary for good current pick-up and for Li-ion mobility. Environmentally benign MOFs are potentially excellent materials for new batteries.

7.8.5 *Ferroelectricity*

Ferroelectrics possess a permanent dipole moment which is reversible in the presence of an applied voltage. For this reason they also exhibit piezoelectric and

Polar point groups

Point groups in which every operation leaves *more* than one point that is not moved. In space groups derived from these point groups, the origin is not fixed in space.

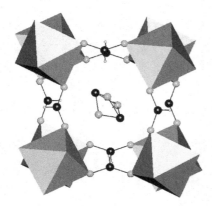

Structure of $(CH_3)_2NH_2)[Zn(HCO_2)_3]$.

Multiferroic materials

Multiferroic materials show two or more ferroic properties simultaneously. A single phase material in which ferroelectricity and magnetic ordering coexist can be used for magnetic data storage and in spintronic devices provided a simple and fast way can be found to turn the electrical and magnetic properties on and off. Some multiferroic materials are $BiFeO_3$, $BiMnO_3$, $TbMnO_3$ and $BaNiF_4$.

pyroelectric properties. In order for a crystalline material to have a permanent dipole it should be belong to one of the ten polar point groups: 1, 2, *m*, *mm*2, 3, 3*m*, 4, 4*mm*, 6 and 6*mm*. The polar state is a consequence of the structural transition from a high-temperature, high-symmetry paraelectric phase to a low-temperature, low-symmetry ferroelectric phase. Well-known ferroelectric materials include KH_2PO_4, $BaTiO_3$, $LiNbO_3$ and Rochelle salt (potassium sodium tartrate tetrahydrate). The later may be considered as a coordination polymer in a broader sense.

Materials that exhibit electrical ordering are of great interest because of their technological importance. $(CH_3)_2NH_2)[Zn(HCO_2)_3]$ illustrates a ferroic transition between paraelectric and antiferroelectric states The coordination polymer anion has the **pcu** topology. The dimethylamine cation occupies the center of each cube and the nitrogen atom is disordered in three different positions. As a consequence, disordered hydrogen bonding exist between the hydrogen atoms NH_2 groups and oxygen atoms from the formate framework (N···O ~2.9 Å). Such ordering of the disordered hydrogen bonding of this type observed here can lead to ferroelectric (e.g., in KDP) or antiferroelectric (as in $NH_4H_2PO_4$) transitions on cooling. As expected this crystal exhibits a phase transition at 156 K due to ordering of the disordered structures of NH_2 hydrogen atoms. A dielectric anomaly expected for a perovskite was found around 160–185 K. This material becomes antiferroelectric below 160 K having a high dielectric constant with an ε_r value of 15.

A multiferroic property has been observed in $[(CH_3)_2NH_2][M(HCO_2)_3]$ (where M = Mn, Fe, Co and Ni). The phase transition between the rhombohedral paraelectric phase to the monoclinic antiferroelectric one involves hydrogen bonded ordering of dimethylammonium cations in the range 160–185 K. On cooling below 8–36 K, antiferroelectric order coexists with weakly ferromagnetic order.

7.8.6 *Birefringence*

Materials exhibit different refractive indices (*n*) in different crystallographic directions and this property is called birefringence. Such materials have a wide range of applications in nonlinear optical processes, liquid-crystal displays, optical filters, and optical components such as quarter-wave plates used to form circularly polarized light. The birefringence magnitude is then defined by $\Delta n = n_e - n_o$; where n_e and n_o are the refractive indices for polarizations parallel (extraordinary) and perpendicular (ordinary) to the axis of anisotropy respectively. Simple inorganic compounds have a high polarizability. For example rutile (TiO_2), a typical birefringent material, with a birefringence (Δn) of 0.29. One can design highly anisotropic coordination polymeric structures using building blocks containing highly polarizable soft atoms that could lead to high Δn. For example, the highly corrugated (4,4) grid polymers [Pb(terpy){M(CN)$_2$}$_2$] (where terpy = 2,2'-6'2''-terpyridine) shows Δn of 0.39 (M = Au) and 0.43 (M = Ag), whereas the corresponding Mn(II) derivative [Mn(Br-terpy){Au(CN)$_2$}$_2$] has an exceptionally high value of 0.50 for Δn.

7.8.7 *Negative Thermal Expansion*

Most solids expand on heating; however a few exhibit negative thermal expansion (NTE) or no expansion due to heating (zero thermal expansion; ZTE). These materials have diverse potential applications in thermal compensation. Of these NTE are very rare and exhibited by coordination polymers. The unprecedented NTE observed in various cyanide bridged coordination polymers such as M(CN)$_2$ (M = Cd, Zn) with two interpenetrating diamond nets is noteworthy. Cd(CN)$_2$, has a linear coefficient of thermal expansion $\alpha = dl/ldT = -20.4 \times 10^{-6}$ K^{-1} over the temperature range 150–375 K. The observed NTE has been attributed to two different modes of transverse motion of the linear cyanide bridge, which draw adjacent metal sites closer together on heating.

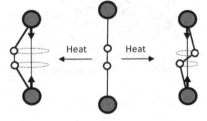

The traverse motion of the linear cyanide bridge in Zn(CN)$_2$.

7.8.8 *Processability*

Unlike organic polymers, coordination polymers are highly crystalline and cannot be processed due to insolubility and decomposition before melting. Different strategies can be used to make coordination polymeric gels, fibers, and different shapes in nano- and microsizes. For example, a thermally responsive blue gel is formed at room temperature when the lipophilic Co(II) complex of 4-(3-lauryloxy) propyl)-1,2,4-triazole is dissolved in chloroform and the blue color of the gel is due to tetrahedral geometry of the Co(II) which shows characteristic absorption around 580–730 nm.

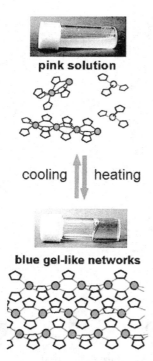

pink solution

cooling ⇅ heating

blue gel-like networks

Formation of 1-D coordination poly-meric gel from coordination complex by heating.

The gel is comprised of the network of fibers with widths of 5–30 nm. When the gel is cooled down to 0°C, it turns into a pale pink solution, suggesting octa-hedral geometry for Co(II). The sol-gel thermochromic transition is reversible upon changing of temperature. For conventional organogels the gel-like network will be formed upon cooling, which dissolve upon heating. These observations indicate that tetrahedral Co(II) monomers are present as low-molecular weight species below 25°C. They are self-assembled to 1-D coordination polymers by heating and form gel-like networks. Upon cooling the solution below 25°C, the tetrahedral Co(II) polymers are converted to monomeric Co(II) complex with octahedral geometry.

Highly fluorescent spherical micro- and nanoparticles composed of 1-D coordination polymers in amorphous state can be synthesized by the coordination-driven assembly of metal ions and homochiral carboxylate-functionalized binaphthyl bis-metallotridentate Schiff base building blocks followed by the fast precipitation with an antisolvent. The particle size can be controlled by the rate of addition and type of initiation solvent used. For instance, rapid addition of diethyl ether pro-duces nanoparticles with a size of ~190 nm. The Zn(II) metal can easily be exchanged with Cu(II), Mn(II) and Pd(II) without affecting the size or morphology of the particles.

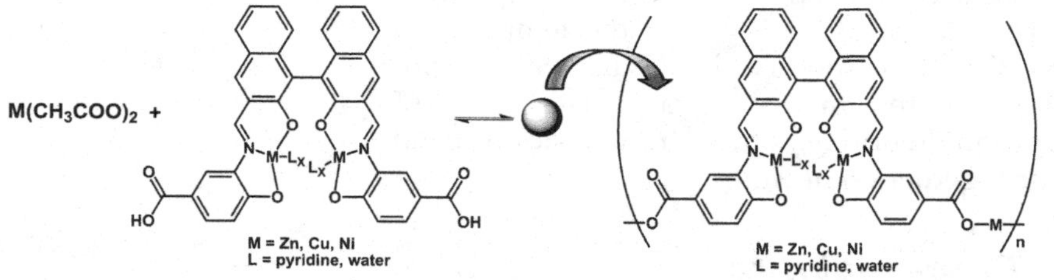

$M(CH_3COO)_2 +$

M = Zn, Cu, Ni
L = pyridine, water

M = Zn, Cu, Ni
L = pyridine, water

Colloidal microspheres formed from 1-D coordination polymers.

Coordination polymers of nano-sized spheres with diameter 58 nm can be synthesized by pouring methanol into a solution containing $TbCl_3$ and di(methyl-ammonium) salt of the anticancer drug disuccinatocisplatin after adjusting the *p*H to 5.5 with dilute NaOH. Successful and reproducible synthesis of NCPs in this system depends on the careful control of the pH of the aqueous precursor solution. These nanospheres can be encapsulated in shells of water soluble polyvinyl-pyrollidone functionalized amorphous silica for the controlled release of the platinum drug.

7.8.9 *Chemical Reactivity*

Solid-state transformations in coordination polymers can be induced by light, heat, solvent/guest removal, uptake or exchange, mechanical stress, etc. The reactivity may result in simple substitution or replacement of a ligand or guest molecule, change in the coordination number or geometry, oxidation state of the metal centers, conformational changes in the structure, functional group replacement at the backbone of the ligand in the structure, dramatic distortion in the framework of the structure, complete change in the dimensionality of the structure, change in the strength and direction of the non-covalent interactions, symmetry of the unit cell, color, optical and magnetic properties. In this process, single crystals may be preserved at the end of the reaction, or the product may render crystalline or amorphous powder.

To determine the changes in the three-dimensional structures in the solids caused by these external stimuli, X-ray crystallography has proved to be an invaluable tool. Therefore reactions in which a single crystal of the reactant transforms to a single crystal of the product (SCSC) are especially amenable to study.

7.8.9.1 *Structural Transformations on Heating*

Here a simple structural transformation induced by temperature is illustrated. A zigzag one-dimensional polymeric structure is observed for the compound [Zn(μ-4,4'-bpy)Cl$_2$] at room temperature. In this polymer Zn(II) has tetrahedral coordination geometry by bonding to two terminal chloride ions and two nitrogen atoms from two different bpy ligands. When the single crystal of this compound is cooled below 130 K, the zigzag structure is converted to 2-D sheet structure as shown. The two terminal chloride ions present in the tetrahedral Zn(II) center in the zigzag polymer now bridge the metal ions in the ladder structure to provide octahedral coordination geometry at Zn(II). The structural transformation is accompanied by an increase of coordination number from 4 (tetrahedral geometry) to 6 (octahedral geometry). This 2-D structure

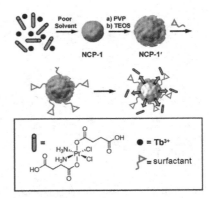

Formation of nanoparticles of 1-D coordination polymer between Pt(II) complex and Tb(III) followed by silica coating and surface functionalization of water solubility. The last step is the release of Pt(II) drug in vitro.

Thermal transformation of a zigzag polymer to a 2-D sheet.

Transformation of a ladder coordination polymer to another ladder structure under UV radiation.

Sliding of a 1-D coordination polymer upon solvent removal.

can be converted back to the zigzag structure by heating the single crystals above 360 K. In other words, this conversion is reversible albeit with a very large hysteresis.

7.8.9.2   *[2+2] Cycloaddition Reactions*

Coordination polymers can be made topochemically reactive by bringing the reactive functional groups in the correct orientation and in close proximity. In Chapter 1 we showed that [2+2] cycloaddition reactions can be carried out in the solid state when Schmidt's geometric and distance criteria are satisfied. In the simple ladder structure, the linear spacer ligands are arranged in parallel orientations between the pairs of Zn(II) ions to form infinite architecture. The distance between the pairs of linear spacers can be controlled by the nature of bridging ligands connecting the metal ions and their ionic radii. One can use a number of linear spacer ligands including *trans*-1,2-bis(4-pyridyl) ethylene (bpe) to construct a simple ladder coordination polymer. The C=C bonds in bpe can be made reactive if they occupy the positions of rails in the polymeric ladder structure. In the 1-D coordination polymer $[\{(F_3CCO_2)(\mu\text{-}O_2CCH_3)Zn\}_2(\mu\text{-bpe})_2]$, the bpe ligands are aligned parallel but slightly slip-stacked. The non-bonding distance of 3.74 Å between the center of the C=C bonds of the bpe pairs is congenial for the photochemical cycloaddition reactions to take place in the solid state. This prediction is in fact confirmed when the compound is irradiated under UV light.

7.8.9.3   *Structural Transformations due to Loss of Solvents*

The loss of coordinated and lattice solvent molecules may induce movements in the coordination polymers. In the linear coordination polymer $[Ag(\mu\text{-bpe})(H_2O)]$ $(CF_3CO_2)\cdot CH_3CN$, the removal of water and CH_3CN is accompanied by sliding of the linear coordination polymer relative to one another leading to the formation of a ladder structure.

A second example is the topochemical dehydration of $[Co(NCS)_2(4,4'\text{-bpy})(H_2O)_2]\cdot(4,4'\text{-bpy})$. This is a linear 1-D coordination polymer consisting of $[Co(4,4'\text{-bpy})]$ as the repeating unit. The metal center has octahedral geometry with two thiocyanate and two aqua ligands along with two pyridyl groups from the 4,4′-bpy ligands. Similar ligands are mutually bonded in a *trans* fashion. A non-coordinated 4,4′-bpy molecules present in the lattice is placed between two polymer strands. Each N atom of the free 4,4′-bpy molecule is hydrogen bonded to one of the hydrogen atoms of the aqua ligand to produce a 2-D structure. If the water molecule is removed by thermal dehydration or under reduced pressure, a new bond can be formed between Co(II) and 4,4′-bpy guest molecule. This will lead to the transformation of the hydrogen bonded 1-D linear coordination polymer to a 2-D coordination polymer with (4,4) grid structure.

7.8.9.4 *Reactivity of Supramolecular Isomers*

Four supramolecular isomers of [Zn{Au(CN)$_2$}$_2$] have been synthesized by changing the counter ions, solvents and reaction conditions such as concentration and *p*H. The connectivity of these 3-D coordination architectures involves tetrahedral zinc(II) node and Au(CN)$_2^-$ anion as spacer ligand. The α-form has sixfold interpenetrating quartz topology, whereas the β-form has a five-fold interpenetrated cristobalite network structure. The γ-form is very similar to that of β-form but with four-fold interpenetration. On the other hand δ-form has a three-fold interpenetrated structure. The short Au···Au interactions in the range 3.11–3.33 Å provide additional stability to these interpenetrated networks. The first three polymorphs are luminescent, having solid-state emissions with wavelengths ranging from 390 to 480 nm and the emission energy can be linearly correlated to the Au···Au distance.

The reactivity of α and δ forms towards ammonia gas is different from β and γ forms. The α and δ forms react with NH$_3$ in a step-wise manner yielding [Zn(NH$_3$)$_2${Au(CN)$_2$}$_2$] and [Zn(NH$_3$)$_4${Au(CN)$_2$}$_2$]. These reactions are reversible and the same isomers are regenerated. On the contrary the β and γ forms yield [Zn(NH$_3$)$_4${Au(CN)$_2$}$_2$] which on desorption gives a mixture of α and δ forms. The binding of NH$_3$ to the Zn(II) center alters the Au···Au distances and hence the emission properties. An excess of ammonia produces a white powder, [Zn(NH$_3$)$_4${Au(CN)$_2$}$_2$] which has a single emission peak at 430 nm on excitation band at 365 nm. Removal of this white powder

Structural transformation by dehydration.

from the NH_3 atmosphere produced a yellow powder $[Zn(NH_3)_2\{Au(CN)_2\}_2]$ which shows a single emission at 500 nm with the excitation band at 400 nm. The reversibility of reaction, fast response to ammonia, color change from white to yellow under normal light and blue to green under UV light, and detection limit to 1 ppb make these coordination polymers ideal ammonia sensors.

7.9   Building Approach: Influence of Experimental Conditions

When working with coordination polymers, the crystal engineer has the luxury of making use of the unique properties of metal ions, reversible binding of the ligands, and in having the strong M–L coordination bond as a synthon. Almost all the elements in the periodic table are available for the construction of coordination polymers to incorporate various physical and chemical properties for specific applications. Each metal ion behaves differently and furnishes different types of dimensionalities and topological architectures. A given metal ion can have variable coordination number, coordination geometry and oxidation state and this, in turn, will influence the network structure. Additionally the nature of the ligands and counter ions will also affect the structure of the coordination polymer. Depending on the coordinating ability of anions such as Cl^-, Br^-, I^-, SCN^-, N_3^-, $CH_3CO_2^-$ and NO_3^-, the bonding sites of the metal ions will be blocked and this would change the dimensionality of the polymer. Further, semi-coordinating and non-coordinating anions such as $CF_3SO_3^-$, BF_4^-, ClO_4^-, PF_6^-, SbF_6^- and BPh_4^- may act as templates and direct the structure.

Crystallization is a kinetic process and hence the experimental conditions have greater influence on the formation of coordination network structures. Solvents, concentration, pH, temperature and time will determine the outcome of crystallization. During crystallization the ligands bind to metal ions reversibly and the least soluble polymer will crystallize first irrespective of the metal-ligand ratio used. The ratio in the product may not be the same as that used in the synthesis. All this is influenced by the solvents and concentration. Guest molecules or impurities present in the reaction medium

Thermodynamics versus kinetics

Kinetic and thermodynamic stabilities should not be confused. The former refers to the speed at which equilibrium conditions are reached and the latter refers to concentrations of complex species and ligands at equilibrium. Complexes that react quickly are called labile and those that react slowly are called inert or nonlabile. In general, the second- and third row transition elements usually form kinetically inert complexes. This is due to high LFSE and strength of M–L bonding. First row transition elements, except those with d^3 and low-spin d^6 configurations, and main group metals usually form kinetically labile complexes. Chelation will make the complex less labile. The degree of lability or inertness of the transition metal complex can be correlated with the d electron configuration of the metal ion. Hence d^3, low-spin d^4, d^5 and d^6 metal complexes are inert while the others are labile.

can also act as templates to form a particular structure.

The student will appreciate that there are many factors that play a part in deciding the outcome of an experiment designed to make a coordination polymer. A particular metal ion taken with a particular ligand may yield several alternative possibilities: these could be variations with respect to metal-to-ligand ratio, dimensionality and connectivity. In part, this variability arises from the fact that kinetics and thermodynamics are so evenly matched in the crystallization of coordination polymers. This gives a good handle to the crystal engineer to direct the course of crystallization appropriately. For example, thermodynamic products could be obtained by using hydrothermal methods while kinetic products could be achieved by variations in solvent and concentration. The field of coordination polymers is by far one of the more popular areas of crystal engineering. It is at the threshold of a stage where more deliberate attempts at structural *control* are likely to bear fruit.

7.10 Summary

- A coordination polymer consists of repeating units of coordination complexes connected by exodentate organic ligands to form a structure that is polymeric, at least in one dimension.
- The dimensionality of a coordination polymer can be viewed in terms of molecularity in the other dimensions.
- The connectivity of coordination polymers can be described in terms of network topologies of well known extended inorganic solids.
- Building blocks in coordination polymers can be connected in different ways and can give rise to different network structures. This is called supramolecular isomerism.
- The coordinating ability and size of the anions as well as the nature of the metal ions (coordination number, geometry, oxidation state) available in the toolbox of the crystal engineer can be judiciously used in the design of a coordination polymer.
- Crystallization is a kinetic process and hence the experimental conditions employed for coordination polymers affect considerably features like metal-to-ligand ratio, dimensionality and topology.
- Coordination polymers have the usual tendency to maximize packing density. Hence the void space within their structures is generally minimized by guest inclusion or by interpenetration.
- Interpenetration is the phenomenon in which two or more nets occur in a structure such that they are not directly connected.
- Guest molecules in voids can be exchanged with other guests. Voids are used for gas storage and catalysis.

- During guest exchange, there may or may not be changes in the crystalline nature of the coordination polymer, its single crystal nature and the network topology.
- Due to the modular nature of a coordination polymer, its structure lends itself to specific design for functional applications.
- Postsynthetic modification is the strategy by which new chemical functionality can be conferred to the framework by a suitable modification of the ligands *after* formation of the MOFs. This chemical modification of the structure takes place without affecting its basic framework.

7.11   Further Reading

Books

A. F. Wells, *Structural Inorganic Chemistry*, 1945 (5th edition, 1984).

A. F. Wells, *Three Dimensional Nets and Polyhedra*, 1977.

L. Öhrström and K. Larsson, *Molecule Based Materials. The Structural Network Approach*, 2005.

P. A. Wright, *Microporous Framework Solids*, 2008.

S. R. Batten, S. M. Neville and D. R. Turner, *Coordination Polymers: Design, Analysis and Application*, 2009.

L. R. MacGillivray (ed), *Metal-organic Frameworks. Design and Application*, 2010.

Papers

M. Fujita, Y. J. Kwon, S. Washizu and K. Ogura, Preparation, clathration ability, and catalysis of a two-dimensional square network material composed of cadmium(II) and 4,4′-bipyridine, *J. Am. Chem. Soc.*, 116, 1151–1152, 1994.

O. M. Yaghi, G. Li and H. Li, Selective binding and removal of guests in a microporous metal-organic framework, *Nature*, 378, 703–706, 1995.

T. L. Hennigar, D. C. MacQuarrie, P. Losier, R. D. Rogers and M. J. Zaworotko, Supramolecular isomerism in coordination polymers: Conformational freedom of ligands in [{Co(NO$_3$)$_2$(1,2-bis(4-pyridyl) ethane)$_{1.5}$)}$_n$], *Angew. Chem. Int. Ed.*, 36, 972–973, 1997.

M. Kondo, T. Yoshitomi, K. Seki, H. Matsuzaka and S. Kitagawa, Three-dimensional framework with channeling cavities for small molecules: {[M$_2$(4,4′-bpy)$_3$(NO$_3$)$_4$]·xH$_2$O}$_n$ (M = Co, Ni, Zn), *Angew. Chem., Int. Ed.*, 36, 1725–1727, 1997.

S. R. Batten and R. Robson, Interpenetrating nets: Ordered, periodic entanglement, *Angew. Chem. Int. Ed.*, 37, 1460–1494, 1998.

S. S.-Y. Chui, S. M.-F. Lo, J. P. H. Charmant, A. G. Orpen and I. D. Williams, *Science*, 283, 1148–1150, 1999.

H. Li, M. Eddaoudi, M. O'Keeffe and O. M. Yaghi, Design and synthesis of an exceptionally stable and highly porous metal-organic framework, *Nature*, 276–279, 1999.

R. Robson, A net-based approach to coordination polymers, *J. Chem. Soc., Dalton Trans.*, 3735–3744, 2000.

J. S. Seo, D. Whang, H. Lee, S. I. Jun, J. Oh, J. Jeon and K. Kim, A homochiral metal-organic porous material for enantioselective separation and catalysis, *Nature*, 404, 982–986, 2000.

B. Moulton and M. J. Zaworotko, From molecules to crystal engineering: Supramolecular isomerism and polymorphism in network solids, *Chem. Rev.*, 101, 1629–1658, 2001.

S. R. Batten, Topology of interpenetration, *CrystEngComm*, 18, 1–7, 2001.

M. Eddaoudi, D. B. Moler, H. Li, B. Chen, T. M. Reineke, M. O'Keeffe and O. M. Yaghi, Modular chemistry: Secondary building units as a basis for the design of highly porous and robust metal-organic carboxylate frameworks, *Acc. Chem. Res.*, 34, 319–330, 2001.

O. R. Evans and W. Lin, Crystal engineering of NLO materials based on metal-organic coordination networks. *Acc. Chem. Res.*, 35, 511–522, 2002.

M. P. Suh, J. W. Ko and H. J. Choi, A metal-organic bilayer open framework with a dynamic component: Single-crystal-to-single-crystal transformations, *J. Am. Chem. Soc.*, 124, 10976–10977, 2002.

O. M. Yaghi, M. O'Keeffe, N. W. Ockwig, H. K. Chae, M. Eddaoudi and J. Kim, Reticular synthesis and the design of new materials, *Nature*, 423, 705–714, 2003.

S. Kitagawa, R. Kitaura and S. Noro, Functional porous coordination polymers, *Angew. Chem. Int. Ed.*, 43, 2334–2375, 2004.

S. Kitagawa and K. Uemura, Dynamic porous properties of coordination polymers inspired by hydrogen bonds, *Chem. Soc. Rev.*, 34, 109–119, 2005.

N. L. Toh, M. Nagarathinam and J. J. Vittal, Topochemical photodimerization in the coordination polymer of [{(CF$_3$CO$_2$)(μ-O$_2$CCH$_3$)Zn}$_2$(μ-bpe)$_2$]$_n$ through single-crystal transformation, *Angew. Chem. Int. Ed.*, 44, 2237–2241, 2005.

G. Feréy, C. Mellot-Draznieks, C. Serre and F. Millange, Crystallized frameworks with giant pores: Are there limits to the possible? *Acc. Chem. Res.*, 38, 217–225, 2005.

J. J. Vittal, Supramolecular structural transformations involving coordination polymers in the solid-state, *Coord. Chem. Rev.*, 251, 1781–1795, 2007.

X.-M. Chen and M.-L. Tong, Solvothermal in situ metal/ligand reactions: A new bridge between coordination chemistry and organic synthetic chemistry, *Acc. Chem. Res.*, 40, 162–170, 2007.

R. Robson, Design and its limitations in the construction of bi- and poly-nuclear coordination complexes and coordination polymers (aka MOFs): A personal view, *Dalton Trans.*, 38, 5113–5131, 2008.

G. Feréy, Hybrid porous solids: Past, present, future, *Chem. Soc. Rev.*, 37, 191–214, 2008.

J. R. Long and O. M. Yaghi, The pervasive chemistry of metal-organic frameworks, *Chem. Soc. Rev.*, 38, 1213–1214, 2009. The student should read all the papers in this special issue.

Z. Wang and S. M. Cohen, Postsynthetic modification of metal-organic frameworks, *Chem. Soc. Rev.*, 38, 1315–1329, 2009.

V. A. Blatov, M. O'Keeffe and D. M. Proserpio, Vertex-, face-, point- Schläfli-, and Delaney-symbols in nets, polyhedral and tilings: Recommended terminology, *CrystEngComm.*, 12, 44–48, 2010.

W. L. Leong and J. J. Vittal, One-dimensional coordination polymers: Complexity and diversity in structures, properties, and applications, *Chem. Rev.*, 111, 688–764, 2011.

7.12 Problems

1. Give two examples from the literature for each of the following: (i) Neutral linear coordination polymers; (ii) Helical coordination polymeric cations. (iii) Cationic coordination polymers with square sheets; (iii) Cationic coordination polymers containing hexagonal sheets.

2. What are the Wells symbols for the five Platonic solids: tetrahedron, cube, octahedron, icosahedron and dodecahedron.

3. What are the Wells extended point symbols for the following topologies: (i) **dia** (ii) **cds** (iii) **pts** (iv) **ptt**.

4. Suggest combinations of metal ions and ligands that could be used to make coordination polymers with the Pt$_3$O$_4$ and CaF$_2$ topologies. See the table in the text.

5. The MOF [Co$_2$(bptc)(H$_2$O)(DMF)$_2$] with the PtS structure was obtained as a kinetic product. On prolonged heating, the same reactants yielded [Co$_2$(bptc)(H$_2$O)$_5$] with the NbO topology. Search the literature to find more information, draw both the

structures and comment on the formation of these structures. The structure of the bptc ligand is given below.

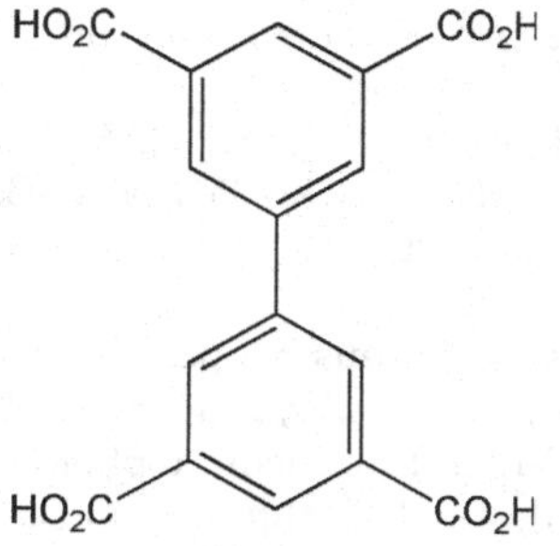

6. Consider the following tetradentate ligand, L:

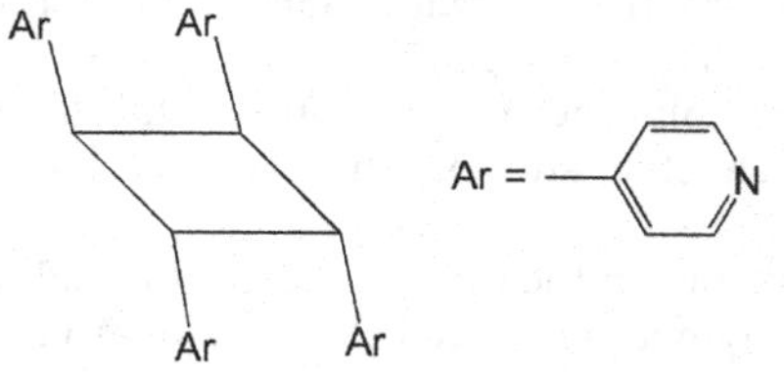

(i)   Suggest a method for the preparation of this ligand, using information in Chapter 6 on topochemical photodimerization of engineered co-crystals.

(ii) Write down the structure of the coordination polymer, [{Cu$_2$(CH$_3$CO$_2$)$_4$}$_2$L] that is obtained when this ligand is treated with Cu(CH$_3$CO$_2$)$_2$.

7. When NiCl$_2$ and pyrazine-2-carboxylic acid (L) are recrystallized under ambient conditions, three distinct solids with compositions NiL$_2$(H$_2$O)$_2$, NiL$_2$(H$_2$O)$_2$ and NiL$_2$ appear concomitantly. All three structures are based on nickel pyrazine carboxylate. The first is a discrete complex connected with non-bonding interactions. The second and third are 1-D coordination polymers. (i) Write down the structures of the three compounds; (ii) Rationalize the connectivity based on crystal structure and identify the coordination and nonbonding interactions; (iii) Are these compounds polymorphs, pseudopolymorphs or supramolecular isomers? (iv) List the techniques that allow you to distinguish these structures experimentally.

8. CoCl$_2$ forms a 2-D coordination polymer with pyrazine, Co(pyz)$_2$Cl$_2$ in which cobalt is octahedral. Write down the structure of this compound. Suggest other structures *with the same composition* that might be formed with pyrazine by ions like Ni(II), Cu(I), Cu(II) and Zn(II). Suggest possible reasons for the formation or otherwise of these alternative structures.

9. Write down the structures of the following 1-D coordination polymers: (i) [Co(4,4'-bpy)(μ$_2$-O$_2$CCH$_3$)$_2$]; (ii) [Co(4,4'-bpy)$_{2.5}$(NO$_3$)$_2$]; (iii) [Co(4,4'-bpy)$_{1.5}$(NO$_3$)$_2$]; (iv) [Co(4,4'-bpethy)$_{1.5}$(NO$_3$)$_2$]·bpy = 4,4'-bipyridine and bpethy = 1,2-*bis*(4-pyridyl)ethyne. One among the third and fourth structure above is interpenetrated. Identify this structure and give a reason.

10. Identify the metal-ligand (ML) stoichiometries for each of the following networks.

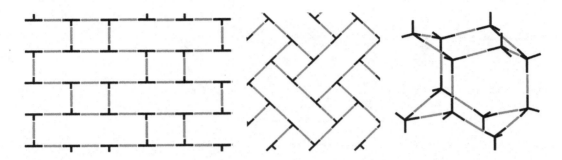

11. When does a 1-D coordination polymer entangle? What are the topologies of the entangled structures?

12. Why are 1-D and 2-D coordination polymers not used for gas storage?

13. The transformation of the linear polymer [Ag(μ-bpe)(H$_2$O)](CF$_3$CO$_2$)·CH$_3$CN to a ladder structure is accompanied by loss of single crystal character in the product. Give a reason for this observation. bpe = *trans*-1,2-bis(4-pyridyl)ethylene.

14. Why are mild reagents and gentle procedures used in postsynthetic modification of MOFs?

Glossary

Acicular Needle like.

Agglomerate A cluster of crystals joined in an irregular fashion to form a material.

Amorphous A solid material that is not crystalline. Amorphous solids lack the long range internal ordering that is characteristic of crystals.

Anisotropic crystal A crystal that exhibits different properties when tested along axes in different directions. In an anisotropic crystal, the interactions in different directions are distinct in terms of strength, isotropy and distance dependence. The crystal anisotropy may pertain to crystal shape/morphology, hardness/softness, reactivity, refractive index, conductivity and in general, to any physical or chemical property.

Antisolvent A solvent in which a compound is not soluble and which is added to a solution of the compound in another solvent, so that crystals of the desired compound precipitate.

API or **Active pharmaceutical ingredient** The substance in a pharmaceutical drug or a pesticide that is biologically active.

Apohost A host framework structure that is left undisturbed upon removal of a guest species.

Asymmetric unit The smallest collection of atoms from which may be generated the contents of the entire *unit cell*, with the help of the symmetry operations of the crystal. In a molecular crystal, Z′ is the number of molecules in the asymmetric unit. Z′ can be an integer or a fraction. Usually, Z′ = 1. For many symmetrical molecules Z′ = 0.5 so that the full molecule can be generated from half the molecule (asymmetric unit) by the application of a suitable symmetry operation (inversion, two-fold axis, mirror plane).

Bifurcated It is a term used for *intermolecular interactions*, typically *hydrogen bonds*. In a bifurcated hydrogen bond, a single acceptor is approached by two donors, or a single donor is approached by two acceptors.

Binodal A *coordination polymer* that contains two topologically different nodes.

Birefringence The property in which the refractive indices are different along different crystal directions. Calcite is a classical example.

Bragg's law The relationship that connects the angle of diffraction of a beam of electrons/neutrons/X-rays of a particular wavelength with the spacings of a specific plane from which such diffraction occurs. The law takes the form $n\lambda = 2d\sin\theta$.

Bravais lattice   The fourteen possible three-dimensional lattices that can be created using lattice points.

Building block   In the context of *coordination polymers*, it is a neutral or charged molecular repeating unit that forms an extended structure.

Cartesian coordinates   Orthonormal coordinates.

Catemer or **catamer**   An infinite one-dimensional pattern of interactions, usually *hydrogen bonds* that is seen in the crystal structures of carboxylic acids and other compounds. It is an alternative packing possibility to dimer formation, which results in a zero-dimensional pattern.

CCD   Charge Coupled Device.

CCDC   Cambridge Crystallographic Data Centre.

Charge transfer complex   See donor-acceptor complex.

Chirality   The property of a molecule wherein it cannot be superimposed on its mirror image.

Chloro rule   It states that a planar aromatic molecule that has dichloro-substitution adopts a crystal structure in which the shortest axis is around 4 Å.

Chloro-methyl rule or **chloro-methyl exchange rule**   It states that molecules that differ in that a chloro group is replaced by a methyl group will have the same crystal structure. Other variants of this rule are the CH_3-CF_3 exchange rule and the benzene-thiophene exchange rule.

CIF   Crystallographic Information File.

Clathrate   *Multi-component crystal* in which one of the molecules is trapped and is enclosed by a framework of another. It is a type of host-guest compound. The term is used with particular reference to hydroquinone, urea and thiourea. Gas hydrates also fall into this category of crystals.

Close packing   The packing of molecules in a crystal that makes the best use of space in that it minimizes the amount of empty space. Uniform spheres can close pack so that 74% of the available space is used. Close packing implies that any molecule in a crystal strives towards the maximum number of near neighbors. To the extent that an organic molecule is generally not spherically shaped, the crystal symmetry of a close packed organic crystal is much lower than the hexagonal and cubic symmetries that are obtained in the *hcp* and *ccp* system of close-packed spheres.

Co-crystal or **cocrystal**   It is typically a *multi-component molecular crystal* in which the different molecules are linked with hydrogen bonds. Generally, the two or more components in a co-crystal are solids themselves under ordinary ambient conditions. There is a great deal of current discussion on the exact definition of this term, and no clear consensus on what it means, or if the term even lends itself to a precise scientific definition.

Co-former or **coformer**   It is the compound that is selected for co-crystallization with a given compound so that a co-crystal is formed. For co-crystals of drugs the co-former is generally chosen from the GRAS list.

Conformational isomorphism   The occurrence of different conformers in the same crystal structure.

Conformational polymorphism   The occurrence of different conformers in different crystal forms of a substance.

Conglomerate   A mixture of chiral crystals that is obtained by crystallization of a racemic compound.

Coordination complex or **coordination compound**   Metal ion with attached ligands.

Coordination number   The number of nearest neighbors to a molecule in a crystal. For most organic crystals this number is 12. In a metal complex, it is the total number of atoms bonded to the central metal atom/ion.

Coordination polymer   Extended structure of metal ions (M) and organic ligands (L) containing unbroken –M–L–M–L– chains, in one, two and three dimensions.

Crystal   A solid form of a substance with a definite and periodic internal structure. It has an external shape that is made up of symmetrically arranged plane surfaces. A crystal is a solid that gives a discrete X-ray diffraction diagram. In the chemical context, a crystal is a condensed phase in which atoms, ions or molecules are brought together in some degree of order. A crystal is a manifestation of mutual recognition and the result of an ordering process of molecules in which enthalpic factors gain over the entropy dominated situation in solution or in the melt.

Crystal coordinates or **fractional coordinates**   Positions of a point in terms of fractions of the unit cell edges in an axial system that is coincident with the lengths and angles of the unit cell. For example, the long diagonal of the unit cell parallelepiped is the line connecting the points (0,0,0) and (1,1,1).

Crystal Engineering   The understanding of intermolecular interactions in the context of crystal packing and the utilization of such understanding in the design of new solids with desired physical and chemical properties. In the context of *supramolecular chemistry*, crystal engineering is the synthesis of solid state supramolecular structures.

Crystal face   A plane surface that is an external boundary of crystal. It is oriented parallel to planes of molecules in the crystal structure. The symmetry that relates the constituents of a crystal also relates the crystal faces.

Crystal habit   The general shape of a crystal. The crystal habit is determined both by the internal structure and the environment in which the crystal grows. Natural crystals rarely grow in ideal geometric shapes. Nevertheless the angular relationships between crystal faces always provide evidence of the symmetry relationships between crystal faces.

Acicular: needle-like
Fibrous: very thin, hair-like
Prismatic: longer than wide, surrounded by parallel faces
Platy, lamellar: thin sheets
Tabular: like a tablet of paper
Equant: nearly equal in all three dimensions

Crystallization   The process in which a crystalline substance is formed from a homogeneous or a heterogeneous medium. These media include solids, liquids or gases, or mixtures thereof that contain one or more chemical substances. The most common instances of crystallization are from a solution or from a melt.

Crystal structure   The arrangement of atoms, molecules, or ions that constitutes the internal structure of a crystal. In the context of crystal engineering, crystals are ordered supramolecular systems, and crystallization is an impressive display of periodic molecular assembly.

Crystal structure prediction or **CSP**   The *ab initio* computer prediction of the crystal structure of an organic or metal-organic molecule, given only the molecular structure.

Crystal systems   Unique systems with distinct symmetry relationships between the three lengths and three angles that characterize the crystal unit cell. The cubic, tetragonal, rhombohedral, hexagonal, orthorhombic, monoclinic and triclinic arrangements of lattice points define the seven crystal systems. In turn, these lead to the 14 *Bravais lattices*, 32 crystal classes and 230 space groups.

CSD or **Cambridge Structural Database**   A computer readable library that contains data on more than 550,000 crystal structures of organic and organometallic compounds, coordination complexes, coordination polymers and metal-organic frameworks. The data is in the form of unit cell information and the fractional crystal coordinates of all the atoms in the structure.

Data mining   Selecting some information from a large amount of computer readable crystal data (unit cells, positional parameters of atoms) in order that some focussed scientific conclusion may be drawn.

Deliquescent or **hygroscopic**   Absorbs moisture from the air and becomes liquid.

Diamagnetism   A phenomenon in which induced magnetism in a substance is in the opposite direction to that of the applied magnetic field.

Diamondoid network or **diamond network**   An arrangement of tetrafunctionalized molecules with or without tetrahedrally coordinated metal ions and other ligands that is topologically equivalent to the diamond crystal structure. The molecules and metal ions act as tetrahedral nodes and the intermolecular associations or coordinate bonds act as node connections.

Diffraction   The ability of a wave to bend around the edges of obstacles or holes. The effect is most noticeable when the size of the obstacle or hole is comparable to the wavelength.

Disorder   A situation in a crystal in which equivalent sites in different unit cells are occupied by atoms, groups of atoms or molecules that differ in position, orientation or conformation. Crystallographic disorder can be static or dynamic. In static disorder, these alternative positions, orientations and conformations are time invariant. In dynamic disorder they are. The image obtained in crystallography is a time and space averaged one. Therefore in normal circumstances a disordered structure appears as if it contains atoms, groups of atoms or molecules in an averaged situation.

Donor-Acceptor complex   A *multi-component crystal* made up of an electron rich and an electron deficient component. Generally these components are flat aromatic molecules and the complex is said to be of the π–π type. However, other types of donor-acceptor complexes like the n–σ type are also known.

Efflorescent   Readily loses water from the crystal and becomes anhydrous.

Enantiomers or **optical isomers**   A pair of molecules with non-superimposable mirror image structures. The two forms rotate the plane of polarized light by equal amounts in opposite directions. Enantiomers can crystallize only in the Sohnke space groups.

Enantiotropic polymorphs or **enantiomorphs**   Polymorphs that interconvert at temperatures below the melting temperature.

Entanglement   Formation of ring structures in *coordination polymers* in which at least two rings are interlocked or catenated.

Equilibrium   State of a system in which the macroscopic properties of each phase of the system become uniform and independent of time. A state of thermal equilibrium, hydrostatic equilibrium and chemical equilibrium respectively refers to the temperature, pressure and chemical potential of each component being uniform throughout the system. When all these quantities become uniform, the system is in thermodynamic equilibrium. Solubility equilibrium is an example of complete thermodynamic equilibrium. For example, a saturated solution of a solid in a liquid at a fixed temperature and pressure is in a state of complete thermodynamic equilibrium. If the system is subjected to a small increase in temperature, a small portion of solid will dissolve to restore the equilibrium (if the solubility increases with temperature), while if there is a small decrease in temperature, a small portion of solid will precipitate. This is the basis for determining accurate values of solubility by approaching the equilibrium solubility from both supersaturation and undersaturation directions.

Etter's rules   Empirical rules that have been widely used to predict *hydrogen bond* preferences in organic solids. Several exceptions are known but the rules generally provide a convenient starting point to understand crystal packing in hydrogen bonded solids. The rules state that: (i) All good proton donors and acceptors are used in hydrogen bonding; (ii) Six-membered-ring intramolecular hydrogen bonds form in preference to intermolecular hydrogen bonds; (iii) The best proton donors and acceptors remaining after intramolecular hydrogen bond formation form intermolecular hydrogen bonds to one another.

Eutectic   The point in a composition versus melting point curve for a multicomponent system where the lowest melting temperature is achieved (Greek, *eutectos*, easily melted).

Ferromagnetism   A type of magnetism in which the magnetic moments of atoms in a solid are aligned within domains which can in turn be aligned with each other by a weak magnetic field.

Gibbs free energy (G)   A measure of the driving force for a chemical reaction, typically measured in units of kcal mol^{-1} or kJ mol^{-1}.

GRAS   Generally recognized as safe.

Growth   The agglomeration of matter around a nucleus that increases its size, eventually leading to the formation of a macroscopic crystal.

Halogen bond   A non-bonded interaction between an electrophilic halogen atom and an electronegative species. It is like a hydrogen bond in which a halogen atom plays the role of the electrophile.

Herringbone motif and **π···π interactions**   Recognition between phenyl groups is mediated by interactions that have two extreme geometries: T-shaped edge-to-face herringbone alignment (dipole–quadrupole interaction, energy 1–2 kcal mol^{-1}), and parallel stacking of phenyl rings with face-to-face interaction between π-electron clouds (quadrupole-quadrupole interaction, energy 1 kcal mol^{-1}). Possible variations in these geometries are the vertex-to-face approach, and the presence of offset or slight inclination between the π stacked rings.

High throughput crystallization   A scheme for carrying out a large number of crystallization experiments systematically varying the conditions. These conditions may be solvent, concentration, temperature or, in the case of co-crystals, the co-former structure. High throughput crystallization may be carried out manually or with mechanical assistance (robots). This technique is useful in the detection of new polymorphs, pseudopolymorphs, co-crystals and additionally, in the case of coordination polymers, supramolecular isomers.

Heterosynthon   A *supramolecular synthon* that is constituted with two distinct yet complementary, molecular components. The carboxyl-pyridyl dimer synthon is a common example of a heterosynthon.

Homosynthon   A *supramolecular synthon* that is constituted with two identical but complementary, molecular components. The carboxyl dimer synthon and the primary amide dimer synthon are common examples of homosynthons.

Hydrogen bond   An interaction between a hydrogen atom from a molecule or a molecular fragment X–H in which X is more electronegative than H, and an atom or group of atoms in the same or a different molecule, in which there is evidence of bond formation. It is the most important interaction in *crystal engineering* because it is both strong and directional.

Hydrothermal   The term when used to describe crystallization refers to a process in which an aqueous solution of a sparingly soluble compound is equilibrated at temperatures ranging from boiling to the critical temperature until crystallization takes place.

Insulation or **structural insulation**   When interactions or *synthons* formed by some functional groups in a crystal structure are unaffected by the presence of other functionalities in the molecule.

Intercalation   Reversible inclusion of molecules or groups between layers.

Interdigitation   Insertion of a part of a molecule or extended structure into the empty space in the network of structures.

Interference or **structural interference**   When interactions or *synthons* that would normally be formed by some functional groups in a crystal structure are not observed because of some influence from other functionalities in the molecule. This is the opposite situation from *structural insulation*.

Intermolecular interactions   In the context of *crystal engineering*, this is a blanket term that is applied to all intermolecular stabilization in a crystal. The crystal structure of a molecule is a free energy minimum that results from a balance of attractive and repulsive forces. At equilibrium, this results in stabilizing interactions with varying strengths, directional preferences, and distance dependence properties. Intermolecular interactions in molecular solids are of two types: short- to medium-range isotropic or van der Waals interactions; long-range anisotropic interactions and *hydrogen bonds*. Anisotropic interactions define directional preferences in the mutual recognition of molecules during crystallization. They are electrostatic in nature and operate at long range. These include ionic interactions ($K^+\cdots O^-$), strong and weak *hydrogen bonds* ($O–H\cdots O$, $C–H\cdots O$), and interactions between heteroatoms (halogen$\cdots$halogen).

Interplanar distances   Distances between nearest parallel planes with the same set of *Miller indices*.

Interpenetration   Mutual intergrowth of two or more networks in a structure wherein the networks are in close physical proximity but not linked with covalent or coordinate bonds. A network cannot be separated without breaking a coordinate bond.

Isomers   Molecules with the same molecular formulae but different topological or spatial arrangements of atoms.·

Isomorphism   Near identity of crystal shape, unit-cell dimensions, and crystal packing between substances of similar chemical composition. Ideally, the substances are so similar to each other that they can form a continuous series of *solid solutions*.

Isoreticular   *Metal-organic frameworks* that have the same topology.

Isostructurality   Similarity between crystal structures in terms of cell parameters, packing features, synthon content and chemical information features. Isostructurality is a more subjective criterion of comparison between crystal structures than isomorphism. Isostructural compounds need not have a similar chemical composition, for example naphthalene and anthracene.

Isotropic crystal   A crystal within which the intermolecular interactions are similar in all directions. Rigorously, only cubic crystals are strictly isotropic in that their properties are independent of the direction of testing. Even in lower crystal symmetry systems, crystals may be approximately isotropic and this is manifested in properties such as crystal growth, shape/morphology and hardness/softness being nearly the same in all directions. Naphthalene, sucrose and ice are isotropic crystals.

Kinetic crystal   Metastable crystal that is formed because of some favorable aggregation pattern of molecules.

Lamellar   Plate like.

Landscape or **structural landscape** or **crystal energy landscape**   A collection of crystal structures that express the pathways that molecules may traverse en route to the formation of stable crystal forms. The landscape includes structures of *polymorphs* and perhaps *pseudopolymorphs* along with selected co-crystals. The landscape is an energy profiling of the crystallization event.

Lattice energy   It is the negative of the potential energy required to take molecules with motionless nuclei (0 K) out of a crystal lattice and separate them by an infinite distance, keeping them motionless, without changing the bond angles, bond lengths or dihedral angles of the molecules. The lattice energy thus arises *only* from the intermolecular interactions in a crystal. It is impossible to measure lattice energy experimentally because the molecules are vibrating even at 0 K. The lattice energy is the quantity calculated in crystal packing studies. The lattice energy is to be distinguished from the *sublimation energy*.

Lattice parameters   Three distances and three angles that together characterize the crystal *unit cell*.

Luminescence   The capability of a solid to emit light when subjected to certain conditions. Typical examples are fluorescence, triboluminescence, and thermoluminescence.

Macrocrystal   A crystal large enough to be seen with the naked eye.

Metal-organic framework or **MOF**   A crystalline three-dimensional *coordination polymer* with porosity capable of including guest species.

Metastable state   State of a system in which a perturbation of any one of its defining variables may cause a change to a more stable state. A driving force exists for the transition of a metastable state to a stable state ($\Delta G < 0$ at constant T and p); however the transition does not occur significantly during the time of observation.

Microcrystal   A tiny crystal that cannot be seen as such with the naked eye.

Miller indices   The smallest integers proportional to the reciprocals of the intercepts of a crystal plane on three axes of unit length.

Mineral   A naturally occurring crystalline inorganic substance.

Mixed crystal   Multi-component crystal.

Molecular complex   An association of two or more compounds that has a definite existence in solution.

Molecular crystal or **molecular solid**   A crystal that is comprised of molecules that associate with weak interactions in the range 1–20 kcal mol^{-1}. The vast majority of molecular crystals are those of neutral organic molecules. Crystals of organic ionic compounds are also taken as molecular crystals as a matter of convention.

Monotropic polymorph or **monotrope**   Polymorphs that cannot interconvert at temperatures below the melting temperature.

Motif or Pattern   A group of molecules that form an identifiable pattern in the crystal.

Multi-component crystal or **multi-component molecular crystal**   A crystal that contains more than one chemical constituent. These constituents can be solid, liquid or gas at ambient conditions. This term has been suggested recently and includes all types of co-crystals, solvates, mixed crystals, solid solutions, host-guest compounds, clathrates, intercalates, molecular complexes, organic salts and adducts.

Net or **Network**   An extended ordered structure of nodes and connections between them.

Node or **Vertex**   Connection point in a network structure.

Nucleation   The formation of a critical aggregate of molecules, which may include solvent, that corresponds to the highest energy in the crystallization reaction coordinate.

Organometallic   An organometallic compound is one that contains a metallic element bonded directly to a carbon atom.

Ostwald ripening   Growth of larger crystals at the cost of smaller crystals. The process is favorable because smaller particles have a higher surface energy than larger ones; the higher total Gibbs energy thus gives rise to an apparent higher solubility.

Ostwald's rule of stages   It states that during crystallization what is obtained is not the thermodynamic global minimum structure but rather the nearest metastable form that can be accessed by the system. Crystallization can proceed from one metastable form to the next until the global minimum is achieved, but this is by no means an invariable outcome because some kinetic forms can be indefinitely stable.

Paramagnetism   A type of magnetism characterized by a positive magnetic susceptibility, so that the material becomes weakly magnetized in the direction of an external field. The magnetization disappears when the field in removed.

Pattern   See motif

Pharmaceutical co-crystal   A multi-component crystal in which at least one of the constituents is an active pharmaceutical ingredient.

Phase diagram   Graphical representation of the phases present at chemical equilibrium in a system that contains two or more phases. Phase diagrams may use any pairs of the variables temperature, pressure, and compositions of various phases. Typical examples are temperature–composition and pressure–composition phase diagrams.

Phase rule   Gibbs' phase rule relates the possible stable phases P in an equilibrium system with the number of components, C, and the number of degrees of freedom, F as $F = C + 2 - P$.

Piezoelectric effect   It occurs when voltage is produced between the surfaces of a dielectric solid when a mechanical stress is applied to it. Conversely, when a voltage is applied across certain surfaces of a piezoelectric solid, it undergoes mechanical distortion. This effect is displayed only by noncentrosymmetric crystals. Quartz and Rochelle salt are classical examples.

Polar crystal   A crystal that shows spontaneous polarization parallel to the *polar axis*. There are 10 polar point groups with a unique polar axis among the 21 noncentrosymmetric point groups. For space groups belonging to these 10 point group systems, the origin cannot be fixed uniquely in space. This definition of the term *polar* pertains to structural polarity.

Polar molecule   A molecule within which the resultant positive and negative charges are not coincident. This definition of the term *polar* pertains to electrical polarity.

Polycrystalline   A substance that contains many small crystalline domains that are randomly locked together. The sizes of the domains vary depending on the material and its method of preparation.

Polymorphism   The phenomenon in which the same chemical substance has more than one crystal structure.

Post-synthetic modification   In the context of MOFs, it is the strategy by which new chemical functionality can be conferred to the framework by a suitable modification of the ligands *after* formation of the MOFs. This chemical modification of the structure takes place without altering its basic framework.

Powder   Microcrystalline solid. The term is used to identify a crystalline sample that is not a single crystal.

Pseudopolymorph   Solvated forms of a compound that have different crystal structures and/or differ in the nature of the included solvent. The term has recently been extended to other types of multi-component crystals.

Reductionism   It is an approach to understand the nature of complex things by reducing them to the interactions of their parts, that is to simpler or more fundamental things. Philosophically, it amounts to stating that a complex system is no more than the sum of its parts.

Refinement   In the context of X-ray crystallography, the procedure in which a proposed structural model is compared to the model obtained from the experimental diffraction intensities, with the aim of minimizing the differences between the two models.

Retrosynthesis   A mode of analysis wherein a synthetic target is dissected into smaller fragments which upon assembly, will give the desired material. This strategy is applicable both in traditional organic synthesis and in *crystal engineering*.

Rietveld analysis   A quantitative x-ray powder diffraction phase analysis of crystalline compound and mixtures.

Robustness   There are two quite different and important meanings for this term in crystal engineering: (i) A structure, usually a network, is *robust* if it withstands loss of internal solvent or guest species. This meaning is close to the English language meaning of the word "robust" in the sense of being strong; (ii) A synthon is said to be *robust* if it appears repeatedly in crystal structures of molecules that contain the relevant functional groups. Around 33% of carboxylic acids feature the carboxyl dimer synthon, and this is a measure of synthon robustness.

SCSC   Single crystal to single crystal transformation.

Secondary Building Unit or **SBU**   A polynuclear metal aggregate that defines the geometry of a node.

Seed   A small single crystal that is used to grow larger crystals from a supersaturated solution or melt.

Self assembly   The phenomenon in which many molecules interact via a predefined pattern of directional non-covalent or coordination bonds.

SHG or **second harmonic generation**   A phenomenon in which a crystal is irradiated with light of frequency ω causing a beam of frequency 2ω to emerge. A mandatory condition for SHG is that the crystal should be noncentrosymmetric.

Single crystal   A crystalline solid in which atoms are arranged in a single specific pattern throughout the material.

Solid solution   A solid that is obtained by co-crystallizing two or more solid substances that are completely miscible. The compositional ranges for solid solution formation are variable and can extend to complete mutual solubility in all proportions.

Solubility curve   A graphical presentation of variation of solubility with temperature.

Solvent-drop grinding   A method of making a co-crystal that involves grinding the components together with a few drops of a solvent. Often, co-crystal formation may not take place in the absence of solvent.

Solving a structure   In the context of X-ray crystallography, the procedure by which the (phaseless) structure factors are assigned phases so that an electron density map is produced, which can be examined for chemical reasonableness.

Sublimation energy   It is the energy required to take molecules in the lowest vibrational state in the crystal at 0 K and to separate them by an infinite distance, allowing the isolated molecules to occupy their ground intramolecular vibrational state in the lowest energy conformation, with relaxation of their equilibrium bond lengths and bond angles to their expectation values in the ground vibrational state of the isolated molecule. Note that the sublimation energy is not the same as the *lattice energy* because the conformation of the molecule in the crystal need not be the lowest energy conformation. The heat of sublimation at room temperature is equal to the sum of the heat of fusion at the melting point, the heat of vaporization at the boiling point, and the difference between the heat capacity of the solid and liquid and that of the vapor, integrated from room temperature to the boiling point.

Superconductivity   is the phenomenon in which a solid loses electrical resistance below a certain critical temperature (T_c). A superconducting solid has zero electrical resistance and shows the Meissner effect.

Supersaturation or **supersaturated solution**   It has a greater composition of a solute than one that is in equilibrium with undissolved solute at specified values of temperature and pressure. Crystallization is a non-equilibrium process that occurs from a condition of supersaturation.

Supramolecular chemistry   The chemistry of molecular assemblies, intermolecular bonds, noncovalent interactions, and hydrogen bonds. Supramolecular signifies that which is beyond the molecule. Supramolecular chemistry is concerned with strategies for the controlled organization of multiple components into complex matter.

Synthon or **supramolecular synthon**   A structural unit within a crystal that can be formed and/or assembled by known or conceivable synthetic operations involving intermolecular interactions. A useful synthon occurs often and has a sufficiently simple structure so that it can be assembled easily from well-known functional groups.

Tautomeric polymorphs or **tautomorphs**   Crystal structures, each of which contain different tautomeric structures of a molecule.

Tecton   A molecule that has identifiable points of linkage to other molecules in the process of assembly into a crystal. In the limit, all molecules are tectons, and so there is an element of subjectivity in using this terminology. The tecton-synthon model for crystal engineering lends itself easily to a retrosynthetic depiction.

Thermodynamic crystal   The crystal with the lowest possible free energy in any particular chemical system. It is the crystal which is at the global minimum of the structural landscape for a compound.

Topochemistry   Chemistry that is affected by the mutual orientation of molecules.

Twinning   Intergrowth of two or more crystals of the same compound in a definite orientation determined by its crystal structure. The twin relationship is specified by a plane on which the structures meet and/or an axis about which one structure is rotated relative to the other. The twinning can be related by reflection across that plane.

Uninodal   A *coordination polymer* that contains only one topological type of node.

Unit cell   Parallelepiped that can be repeated infinitely in three dimensions to generate the entire crystal.

van't Hoff equation   An equation that expresses the temperature dependence on the equilibrium constant K (e.g. solubility) of a chemical reaction:

$$\frac{d \ln K}{dT} = \frac{\Delta_r H^\circ}{RT^2}$$

where $\Delta_r H^\circ$ is the standard enthalpy of reaction, R the molar gas constant, and T the temperature.

Vapor pressure   The pressure of a gas in equilibrium with a liquid or a solid at a specified temperature.

X-ray diffraction pattern   It is an interference pattern created by X-rays as they pass through a material. Studying X-ray diffraction patterns gives detailed information on the three-dimensional structure of crystals, surfaces, and atoms.

Some Data on Crystallographic Space Groups

Total	230
Triclinic	2
Monoclinic	13
Orthorhombic	59
Tetragonal	68
Rhombohedral	25
Hexagonal	27
Cubic	36
Centrosymmetric	92

$P\bar{1}$, $P2/m$, $P2_1/m$, $C2/m$, $P2/c$, $P2_1/c$, $C2/c$, $Pmmm$, $Pnnn$, $Pccm$, $Pban$, $Pmma$, $Pnna$, $Pmna$, $Pcca$, $Pbam$, $Pccn$, $Pbcm$, $Pnnm$, $Pmmn$, $Pbcn$, $Pbca$, $Pnma$, $Cmcm$, $Cmca$, $Cmmm$, $Cccm$, $Cmma$, $Ccca$, $Fmmm$, $Fddd$, $Immm$, $Ibam$, $Ibca$, $Imma$, $P4/m$, $P4_2/m$, $P4/n$, $P4_2/n$, $I4/m$, $I4_1/a$, $P4/mmm$, $P4/mcc$, $P4/nbm$, $P4/nnc$, $P4/mbm$, $P4/mnc$, $P4/nmm$, $P4/ncc$, $P4_2/mmc$, $P4_2/mcm$, $P4_2/nbc$, $P4_2/nnm$, $P4_2/mbc$, $P4_2/mnm$, $P4_2/nmc$, $P4_2/ncm$, $I4/mmm$, $I4/mcm$, $I4_1/amd$, $I4_1/acd$, $P\bar{3}$, $R\bar{3}$, $P\bar{3}1m$, $P\bar{3}1c$, $P\bar{3}m1$, $P\bar{3}c1$, $R\bar{3}m$, $R\bar{3}c$, $P6/m$, $P6_3/m$, $P6/mmm$, $P6/mcc$, $P6_3/mcm$, $P6_3/mmc$, $Pm\bar{3}$, $Pn\bar{3}$, $Fm\bar{3}$, $Fd\bar{3}$, $Im\bar{3}$, $Pa\bar{3}$, $Ia\bar{3}$, $Pm\bar{3}m$, $Pn\bar{3}n$, $Pm\bar{3}n$, $Pn\bar{3}m$, $Fm\bar{3}m$, $Fm\bar{3}c$, $Fd\bar{3}m$, $Fd\bar{3}c$, $Im\bar{3}m$, $Ia\bar{3}d$

Noncentrosymmetric	138
Noncentrosymmetric and with mirror planes and/or glides	73

Pm, Pc, Cm, Cc, $Pmm2$, $Pmc2_1$, $Pcc2$, $Pma2$, $Pca2_1$, $Pnc2$, $Pmn2_1$, $Pba2$, $Pna2_1$, $Pnn2$, $Cmm2$, $Cmc2_1$, $Ccc2$, $Amm2$, $Abm2$, $Ama2$, $Aba2$, $Fmm2$, $Fdd2$, $Imm2$, $Iba2$, $Ima2$, $P\bar{4}$, $I\bar{4}$, $P4mm$, $P4bm$, $P4_2cm$, $P4_2nm$, $P4cc$, $P4nc$, $P4_2mc$, $P4_2bc$, $I4mm$, $I4cm$, $I4_1md$, $I4_1cd$, $P\bar{4}2m$, $P\bar{4}2c$, $P\bar{4}2_1m$, $P\bar{4}2_1c$, $P\bar{4}m2$, $P\bar{4}c2$, $P\bar{4}b2$, $P\bar{4}n2$, $I\bar{4}m2$, $I\bar{4}c2$, $I\bar{4}2m$, $I\bar{4}2d$, $P3m1$, $P31m$, $P3c1$, $P31c$, $R3m$, $R3c$, $P\bar{6}$, $P6mm$, $P6cc$, $P6_3cm$, $P6_3mc$, $P\bar{6}m2$, $P\bar{6}c2$, $P\bar{6}2m$, $P\bar{6}2c$, $P\bar{4}3m$, $F\bar{4}3m$, $I\bar{4}3m$, $P\bar{4}3n$, $F\bar{4}3c$, $I\bar{4}3d$

Noncentrosymmetric and chiral (Sohnke space groups)	65

$P1$, $P2$, $P2_1$, $C2$, $P222$, $P222_1$, $P2_12_12$, $P2_12_12_1$, $C222_1$, $C222$, $F222$, $I222$, $I2_12_12_1$, $P4$, $P4_1$, $P4_2$, $P4_3$, $I4$, $I4_1$, $P422$, $P42_12$, $P4_122$, $P4_12_12$,

$P4_222$, $P4_22_12$, $P4_322$, $P4_32_12$, $I422$, $I4_122$, $P3$, $P3_1$, $P3_2$, $R3$, $P312$, $P321$, $P3_112$, $P3_121$, $P3_212$, $P3_221$, $R32$, $P6$, $P6_1$, $P6_5$, $P6_2$, $P6_4$, $P6_3$, $P622$, $P6_122$, $P6_522$, $P6_222$, $P6_422$, $P6_322$, $P23$, $F23$, $I23$, $P2_13$, $I2_13$, $P432$, $P4_232$, $F432$, $F4_132$, $I432$, $P4_332$, $P4_132$, $I4_132$

Enantiomorphic   11 pairs
$P3_1$ ($P3_2$), $P4_1$ ($P4_3$), $P6_1$ ($P6_5$), $P6_222$ ($P6_422$), $P3_112$ ($P3_212$), $P4_122$ ($P4_322$), $P6_2$ ($P6_4$), $P4_132$ ($P4_332$), $P3_121$ ($P3_221$), $P4_12_12$ ($P4_32_12$), $P6_122$ ($P6_522$)

Some additional information (from 537,899 structures in the CSD)

- Noncentrosymmetric space groups account for 23% of all crystal structures.
- All space groups are not equally populated. Almost 35% of all organic, organometallic and metal-organic compounds crystallize in the single monoclinic space group $P2_1/c$. The most populated noncentrosymmetric space group is $P2_12_12_1$ which is populated by 7.7% of all structures. 18 space groups are used less than 10 times each. 136 space groups are used less than 100 times each.
- $P2_1/n$ and $P2_1/a$ are not found in the above list of space groups because of their alternative designation of crystallographic axes. They are equivalent to $P2_1/c$.

List of Useful Web Sites

Keywords	Brief description	Website link
Crystallographic resources	Useful information in the day-to-day work of crystallographers.	http://www.iucr.org/resources
Crystallographic education	Material for teachers and links to external educational web sites.	http://www.iucr.org/education
Space group diagrams	A hypertext book of crystallographic space group diagrams and tables.	http://img.chem.ucl.ac.uk/sgp/mainmenu.htm
checkCIF	Reports on the consistency and integrity of crystal structure determinations reported in CIF format.	http://checkcif.iucr.org/
Crystallographic terminology	Online dictionary of crystallography.	http://reference.iucr.org/dictionary/Main_Page
SHELX97	SHELX-97 Manual.	http://shelx.uni-ac.gwdg.de/SHELX/shelx.pdf
PLATON	Versatile SHELX97 compatible multipurpose crystallographic tool.	http://www.cryst.chem.uu.nl/platon/
Growing single crystals	Tips on growing crystals for X-ray structure analysis.	http://www.xray.ncsu.edu/GrowXtal.html
		http://www.cryst.chem.uu.nl/lutz/growing/growing.html
Crystallographic Fonts	Fonts for symmetry elements and space group symbols for word processing and other programs.	http://www.iucr.org/resources/symmetry-font http://www.x-seed.net/freestuff.html
CCP14	Collaborative Computational Project for powder diffraction and small-molecule single-crystal diffraction.	http://www.ccp14.ac.uk/

(Continued)

(Continued)

Keywords	Brief description	Website link
High Z'	Information on structures with more than one molecule in the asymmetric unit, i.e. Z' > 1.	http://www.dur.ac.uk/zprime/
Cambridge Structural Database (CSD)	Depository of more than 1/2 million small molecule crystal structures	http://www.ccdc.cam.ac.uk/prods/ccs/csd.html http://www.ccdc.cam.ac.uk/free_services/csdsymmetry/
ICDD	Powder diffraction data.	http://www.icdd.com/
ICSD	Inorganic Crystal Structure Database.	http://www.fiz-karlsruhe.de/icsd.html
CRYSTMET	Database for metals, alloys, minerals and intermetallics.	http://www.tothcanada.com
Protein Data Bank (PDB)	Database for proteins, nucleic acids and complex assemblies.	http://www.rcsb.org/pdb/home/home.do
Crystal Growth & Design (CGD)	Crystal engineering journal from American Chemical Society.	http://pubs.acs.org/journal/cgdefu
CrystEngComm	Crystal engineering journal from Royal Society of Chemistry.	http://pubs.rsc.org/en/journals/journalissues/ce
CGD Network	Crystal Growth & Design online forum for discussions and comments in the field.	http://acswebcontent.acs.org/cgdnetwork/index.html
CrystEng-Community	Virtual web community for crystal engineers, providing links to research groups, conferences and events, and highlighting the latest research in the field.	http://www.rsc.org/Publishing/Community/index.asp
xforum	Forum for discussion of X-ray and neutron diffraction, hardware and software.	http://www.x-rayman.co.uk/xforum/
Other links	Links to other useful crystallography sites from University of Maryland.	http://www2.chem.umd.edu/facility/xray/XCC_Links.htm
SINCRIS	Software database for crystallography.	http://ww1.iucr.org/sincris-top/logiciel/

(Continued)

(*Continued*)

Keywords	Brief description	Website link
O'Keeffe's NET	Reticular chemistry structure resource.	http://rcsr.anu.edu.au/home
SYSTRE	A program to analyze periodic nets	http://gavrog.sourceforge.net/
Batten	Interpenetration.	http://www.chem.monash.edu.au/staff/sbatten/interpen/index.html
WINGX	MS-Windows system of programs for solving, refining and analyzing single crystal X-ray diffraction data for small molecules.	http://www.chem.gla.ac.uk/~louis/software/wingx/
OLEX	Imports structural data via a number of crystallographic file formats and generates the extended structure if required, and produces a picture. Useful for network connectivity.	http://www.ccp14.ac.uk/ccp/web-mirrors/lcells/olex/olex_index.htm
CRYSTALMAKER	Crystal and molecular structures diffraction.	http://www.crystalmaker.com/
DIAMOND	Molecular and crystal structure visualization software.	http://www.crystalimpact.com/diamond/
Singapore National Crystal Growing Challenge	Contest for high school and junior college students to grow the biggest single crystals.	http://www.chemistry.nus.edu.sg/events/Community Outreach/ncgc/index.htm
Gordon Research Conference in Crystal Engineering	The latest scientific meeting for crystal engineers	http://www.grc.org/

Some Useful Educational References in Crystal Engineering

B. Kahr, J. K. Chow and M. L. Peterson, Painting crystals, Organic hourglass inclusions — a review of past and recent work and a student experiment, *J. Chem. Ed.*, 71, 584–586, 1994.

A. Martin, Hydrogen bonds involving transition metal centers acting as proton acceptors, *J. Chem. Ed.*, 76, 578–583, 1999.

C. V. K. Sharma, Designing advanced materials as simple as assembling Lego blocks, *J. Chem. Ed.*, 78, 617–622, 2001.

D. K. Smith, A supramolecular approach to medicinal chemistry: Medicine beyond the molecule, *J. Chem. Ed.*, 82, 393–400, 2005.

T. Friščič, T. D. Hamilton, G. S. Papaefstathiou and L. R. MacGillivray, A template-controlled solid-state reaction for the organic chemistry laboratory, *J. Chem. Ed.*, 82, 1679–1681, 2005.

G. M. Battle, F. H. Allen and G. M. Ferrence, Teaching three-dimensional structural chemistry using crystal structure databases. 1. An interactive web-accessible teaching subset of the Cambridge Structural Database, *J. Chem. Ed.*, 87, 809–812, 2010.

G. M. Battle, F. H. Allen and G. M. Ferrence, Teaching three-dimensional structural chemistry using crystal structure databases. 2. Teaching units that utilize an interactive web-accessible subset of the Cambridge Structural Database, *J. Chem. Ed.*, 87, 813–818, 2010.

M. L. Myrick, M. Baranowski and L. T. M. Profeta, An experiment in physical chemistry: Polymorphism and phase stability in acetaminophen (paracetamol), *J. Chem. Ed.*, 87, 842–844, 2010.

W. F. Coleman, Comparing solid, gas phase, and solution structures using the Cambridge Structural Database, *J. Chem. Ed.*, 87, 882–883, 2010.

K. Sumida and J. Arnold, Preparation, characterization, and postsynthetic modification of metal-organic frameworks: Synthetic experiments for an undergraduate laboratory course in inorganic chemistry, *J. Chem. Ed.*, 88, 92–94, 2011.

Index